The Typhoon Shipments

Also by Kevin Klose

I WILL SURVIVE (with Sala Pawlowicz)

THE
*T*YPHOON
*S*HIPMENTS

Kevin Klose and
Philip A. McCombs

W · W · NORTON & COMPANY · INC ·

NEW YORK

For our families.

Contents

*This account and all the characters in it are
purely fictional. The theme was suggested by
newspaper reports of certain alleged bizarre events.
The action takes place during the American war
in Vietnam.*

The Typhoon Shipments

Prologue

Gilmore saw the mourners huddled on the grassy hillside. Even at a distance, he could see heads turn to watch as he gunned the blue sedan up the cemetery's winding roadway and skidded to a noisy halt.

The morning sun cast an even, warm light as he walked up the gentle slope. The sharp wail of a bugle cut through the stillness with the music of "Taps."

An Army lieutenant detached himself from the honor guard of stiffly braced soldiers and strode to meet Gilmore.

"Can I help you?" he asked, pugnaciously thrusting his face upward.

Gilmore flapped open his wallet. A federal badge glinted inside.

"Gilmore, Customs. Is this the funeral of Private Joe DiMalco?"

"That's right," said the lieutenant.

"Then we're going to open that coffin."

"My ass you are."

"Son, if that coffin goes into the ground unopened, you'll get busted all the way down to private and I'll have to answer to the secretary of defense. Now get those nice people out of the way because I'm in a hurry."

The lieutenant wavered. "Why?" he asked. "Why open it?"

"National security."

It took several minutes to interrupt the service and move the group of mourners a dozen yards away.

Then Gilmore bent over the plain green military coffin and worked on its fastenings with a screwdriver while the soldiers and mourners silently watched. An aluminum plaque on the side of the coffin read:

PFC JOSEPH V. DIMALCO 526–66–0591
U.S. MILITARY ASSISTANCE COMMAND—VIETNAM
REMAINS NOT SUITABLE FOR VIEWING

Gilmore had begun prying open the lid when he heard a commotion behind him. He turned. DiMalco's mother had broken away from the group and was rushing at him, her arms flailing and her face twisted.

"Joey!" she cried, pushing around Gilmore to look into the coffin. For a moment, she seemed transfixed as she gazed at the thin, blond man lying inside.

"Where's Joey?" she pleaded, stumbling backward with her hands to her eyes.

Gilmore looked into the coffin, then back at the mother.

"That's right, ma'am," he said softly. "That's not Joey in there."

PART I

Gilmore

1

The small door opened with a click and the attendant grabbed the handle at the bottom of the slab and yanked. The stainless steel sheet rolled easily forward, like a large cabinet drawer, with an efficient rumble of bearings on tracks. A naked body emerged, lying on its back. It was the body from the cemetery. A paper shipping tag was tied to the big toe of the left foot. A towel covered the face so that only a thatch of blond hair, cut medium length, was visible.

The attendant drew back the towel. "He the one?"

Daniel Gilmore struggled to keep control. You couldn't tell by looking at the ruined face.

"Yeah, Jesus, that's the one."

It was Charley Luckett, newly wed, newly dead. You'd know him from the blond hair. You didn't need to see the brown eyes or the slightly fleshy cheeks and full lips. All you needed was that hair. The face itself was not there.

Robert Holt, Gilmore's assistant, was shaking his head. "What a mothering waste."

"What kind of animals are we dealing with?" Gilmore's question hung in the clammy room.

"I don't know, Dan," Holt said. "I swear I don't know."

The attendant dropped the towel back. "Let's ID him if you don't mind, sir," he said.

"No, I don't mind," Gilmore replied. The words were automatic, polite. He felt numb.

They went through the identification procedures quickly, rolling the body's fingers onto an ink pad, transferring the prints to a government ID form and then comparing them with a file card of prints that had come over by special messenger from the Treasury Department.

The attendant studied the prints, then wiped his nose with the back of his hand. "Well, looks like they match all right. I guess he's the one."

Gilmore barely heard the words. A spasm of rage and regret boiled through him. "Jesus, you lose a guy . . . like this . . . it hurts," he managed. "It's a new feeling . . ."

"Yeah, I know," said the attendant equably, as one professional to another. "I've seen some bad ones, I'll tell ya." He pushed the tray back into the wall. Charley Luckett disappeared.

Charley Luckett Eagle Scout was what Gilmore had called Luckett: young, bright, tough, incredibly straight Luckett. Trained in karate, photography, botany, chemistry, marksmanship. Passing marks there. Trained in skepticism, caution, stealth. A failure there.

Charley's weaknesses hadn't seemed important three weeks before when Gilmore had sent Luckett to Saigon to look into a new heroin pipeline that was flooding Washington. He, among the ten agents of Task Force Washington that Gilmore commanded, was the best and the smartest and, with seasoning, would become a star. It looked like a good opportunity for Charley. Gilmore gave it to him. And now Charley Luckett had come back in a box.

"Doc's in his office," the attendant said. "He's got the autopsy." He led the way, shouldering through the doors of the morgue. Walter Reed's endless gleaming marble hallways stretched before them.

Gilmore's thoughts veered between framing an apology to

Luckett and framing an explanation to his own bosses. He forced himself to concentrate on an explanation. They were the living, they would come first. He was scheduled to brief the task force coordinating committee the next day. The members would demand a full accounting. Some would want Gilmore's head in exchange for Luckett. They hadn't allowed Gilmore to take over the task force fight against heroin traffic in the nation's capital so he could send out agents to be killed.

The attendant motioned Gilmore and Holt into a small office. A bulldog lieutenant colonel sat behind a battered metal desk. A plastic sign said he was LTC P G PHILLIPS. A Camel hung from his mouth and dribbled ashes on his white smock. Gilmore looked down at the heavy man in the smock and thought: lard-ass.

"I'll read the autopsy," the colonel said, motioning them into metal chairs. "You got any questions, ask." He opened the report and read:

"This is the body of a man identified as Charles Arthur Luckett, which you and you . . ." He looked up at the two agents, then down again. ". . . have confirmed. He was a United States Customs Service agent. It is the body of a well-nourished, well-developed white male weighing an estimated one hundred seventy pounds and measuring seventy-two inches in length. The head is of normal shape, the limbs are normal and of normal proportion to the body . . ."

Just a normal American body with a normal American wife who doesn't yet know that her husband's unexplained business trip to Saigon has been unexpectedly extended into infinity. Luckett had been married five months. The woman had called . . . how many times? Four? Five? Gilmore couldn't recall. The conversations had been brief. No, no,

Mrs. Luckett, we haven't heard from Charley yet. Maybe he's upcountry and out of touch. And what would it be like when he told her? I'm sorry, Mrs. Luckett, I really am. "There is evidence of substantial external post mortem trauma to the face and facial features," the colonel read. Then he analyzed precisely how the killers had made Luckett's face unrecognizable.

Gilmore remembered Valerie Luckett's voice on the telephone—rich, flexible, a voice that hinted at depths of emotion. She had hidden her fears for Charley well. Recalling how she had sounded in those telephone calls, Gilmore knew Charley had gotten himself a good one, strong and tough. Well, she would have to be stronger and tougher than she had ever been before.

"The neck reveals a deep stab wound," the colonel read. "The trachea reveals a puncture opening one-half-inch wide and this same blow also includes the aortic vessel. Luckett got stabbed right about here." Phillips looked up and pointed to his throat just below the point of the Adam's apple. "This is a professional blow. The knife penetrates both blood and wind at once. Virtually no sound at all, except perhaps a slight sigh, or soft choke. Amateurs generally slit like this." He drew his hand across his throat.

"This is a Special Forces wound, an assassin's wound, a single precise blow, straight in and down. There are no nicks or cuts on the victim's hands, implying he was taken from behind, surprised probably, and didn't have a chance to defend himself. You can imply a yoke attack—the assailant grabs with one arm over the neck or mouth from behind, or perhaps pins the arms, and then drives the blow home."

"What else?" Holt asked.

"The only truly amateurish thing is the attempted disfigurement."

"Why do you say attempted?" asked Gilmore. "It more or less succeeded. You couldn't identify Luckett from a picture."

"It didn't succeed at all," said the colonel. "You identified Mr. Luckett officially in a few moments of fingerprinting. This disfigurement was purely superficial. Teeth are still there, the most positive means of identification. Even a fire generally won't destroy teeth. That's why the Army has been able to identify so many war dead from Vietnam. Much better dental records are kept these days. The thing they should have done with Mr. Luckett was cut off his head and hands. That removes the teeth and fingerprints, just like that." He snapped his fingers.

And there it is, thought Gilmore. But how had Luckett wound up in a coffin meant for another? Was he killed in Saigon and then shipped to the morgue there and mistaken for someone else? Or had Luckett indeed gone upcountry and been killed in the boonies, then shipped back to Saigon with other American war dead? If he was misidentified, where was Joe DiMalco, the kid Luckett was meant to be?

Gilmore didn't know what to think. The people who had mutilated Charley Luckett must be of a different breed. The normal rules didn't apply.

Colonel Phillips gave them a copy of the formal autopsy report and they went out into the heavy humidity of a June midday in Washington.

"God," said Gilmore.

"I agree," said Holt.

Neither had spoken during the sweltering drive to their downtown office on the fourth floor of an old building on Vermont Avenue. The office was furnished in Dingy Bureaucratic. Gilmore flipped on the overhead fluorescent lights as they walked in, then slumped into his squeaky reclining office chair. Holt sat in front of the gray metal desk, glumly fingering a toothpick.

They both knew what had to be done, and neither wanted to start the legwork, the telephone calls that you count by hundreds instead of by tens, the endless paperwork. They would have to call the military, call the Air Force base at Dover, Delaware, because that's where Charley Luckett had landed in the coffin. It was just one of a thousand things to do, checks to make. But now it would all be worse because they had lost Charley. The pressure on them because of the increasing flow of heroin would become even more intense—perhaps even unbearable.

They had succeeded so well for eight months. Gilmore and Holt had quietly taken control of Task Force Washington's ten agents while Freddie Moran, the White House special assistant who had dreamed up the idea of an interdepartmental drug task force and was its director, sank into alcoholic seas.

Together, Gilmore and Holt had kept the task force going, and together they had made a success of it. Its operations led directly to the jailing of three big-time pushers and indictments against four others. It was doing just fine under Gilmore's acting directorship. Just fine until the first packet of Typhoon brand heroin showed up in police custody in early April. By mid-May, there was a flood of the stuff, the purest any pusher had ever seen on the streets of Washington. The law enforcement agencies were in a frenzy. The special task force had started taking fire. It was hinted that interest in its successes and failures was noted at the very top, where questions of law and order were politically important. That was why Gilmore had sent Luckett to Saigon, covertly, without advance warning to the agencies there, informing only the highest levels of the embassy.

Now that decision was haunting him, threatening both his own control of the task force and the task force itself. It all fit together, a series of collapsing boxes. Gilmore felt as if he were in the smallest of the boxes, way on the inside, trapped.

For more than a month, Gilmore's attention had centered on a man named Nick Westley, one of Washington's principal drug dealers. Westley was under surveillance, and his photo had been taken perhaps two dozen times, his voice recorded by wiretaps as many times. Big Nick was a large, handsome black man who wore perfectly tailored clothes, who drove a white Mark IV Continental, who had no visible means of support. The wiretaps were frustrating: never more than a few monosyllables at a time. Nick apparently knew his telephone was tapped. He had never been arrested for anything more than a parking violation.

Gilmore's main source of information about Nick, as well as about several other major drug figures, was another black man, an informer about whom Gilmore knew next to nothing but who he supposed was a member of Nick's ring. The informer had provided valuable tips that the task force had acted on swiftly and brutally, netting several drug pushers and tens of thousands of dollars worth of dope.

During the last six weeks, the tips about Nick had multiplied. Nick's presence in town after a long absence roughly coincided with the increased flood of Typhoon heroin. The tips about him were mostly vague—he had his "crew" back together, maybe three or four enforcers and some couriers; the slayings of several small drug dealers were attributed to him ("drug-related execution-style slayings" was the newspaper shorthand for these). The pigeon spoke of a "big, big connection" in the offing.

"Multi-kilo is the talk," he had told Gilmore, and Gilmore had been very excited: Typhoon heroin, direct to you from the Golden Triangle, via Cambodia, the Parrot's Beak, and points west. Pure, uncut gold, pure enough so that if you rub it into your eyelids you can get a high. Eyelids and assholes—those are the favorites. Any soft tissue. No needle freaks need apply; just a quick application of this pure white powder and warts, cuts, headaches, and nightmares go away. Guaranteed to cure just about anything that ails you.

But the pigeon had been vague. He said he was still searching for information. Gilmore was offering him two thousand dollars, maybe even more, for information that would bust Nick, but the pigeon said to hold tight and wait, and he kept taking his weekly hundred-dollar retainer.

Only one tip had been solid. The pigeon didn't think there was heroin involved, but he thought Gilmore should check a certain coffin of a soldier named DiMalco, now dead. The pigeon had heard something, even he wasn't sure just what—a rumor, a whisper. Finding Luckett had been just that easy. It had cost a hundred bucks.

Gilmore had a tough job, but he had help. He had Holt, an old-time fullback of a man with salt-and-pepper hair, unkempt dark blue three-button suit, white shirt from Robert Hall, dark tie, and a friendly, open face, unremarkable in any way. He was solid, functional like a truck fender or a ship's anchor. Holt was forty-nine, married, and the father of three children. He was the stuff of cop movies—utterly competent, thorough, with a vacuum cleaner mind, and an even-tempered curiosity about the world.

And then Gilmore also had his own mind and his own tempering, a peculiar blend of familial pressures and personal drive. In many ways, though twenty years younger than Holt, Gilmore realized that he had seen as much combat. He carried more scar tissue than anyone could guess at.

That was the way you were raised in the Gilmore clan—keep your lumps to yourself and learn from them. It was cruel if you were weak, a tonic if strong. He had suffered in some ways, benefited in others. But he was durable and independent and he knew it.

Life for males in the Gilmore family traditionally followed a set course: six years at any one of three prep schools, four years at Harvard (letters in two varsity sports), two years as a naval deck officer (more in wartime), three years at Harvard Law, and then six decades in one of those multiple-named law firms located in the highrises of Manhattan. Get in deep with both hands and start working. Pile up the salary, the bonus, the partner's share, double the portfolio every five years. Get some directorships and serve on the boards of one or two charities. Pick a wife somewhere along the line. Procreate. Administrate. Hold your liquor. Screw only your wife, but if that isn't enough, don't get caught. Take no vacation before you are thirty-five, take one a year after that. Never retire. Make them carry you out feet first, claimed by cardiac arrest.

Do it right and you've got eighty good years, start to finish. Eight decades of impatient, safe, living death.

But somewhere between childhood and manhood, something changed in Daniel Gilmore; there was some alteration of the chemicals that brought yearning for more than he was getting at one of those three prep schools the family had so generously endowed.

He enlisted in the Army when he was eighteen, angry at things for which he didn't even have a name. He was assigned to military police and shipped to Tokyo. When he was mustered out at age twenty-one, he weighed two hundred pounds and had added another three inches to his six-foot frame. He flew back as the first wave of Vietnam-bound troops was moving across the Pacific. Gilmore applied for college and was accepted at Dartmouth. He found it dull and pointless and after two years of bad grades, he quit and on a whim went to work in Washington for the Customs Service as a baggage inspector. From there he rose quickly.

Shortly after he turned twenty-five, and after being away from the family for almost a decade, Gilmore received a summons from his father for an appearance to discuss his finances and his inheritance.

He went home to the Connecticut country house for an audience with his father and showed him the .38 Special and the badge. They retired into the inner den. The old man poured himself a brandy and soda and sat in his deep leather chair behind the mahogany desk, sipping the drink and fingering the pistol.

At length he said, "I've raised up a goddam cop? A cop?" The words were like acid.

Gilmore stood up, carefully took back the gun and the badge, then leaned toward the old man's stony face and said, "You're goddam right."

It was a declaration and an assertion of self-esteem. High stakes, violence, and risk.

Now, sitting with Holt, Gilmore broke the silence. "We better get going, Bob, dammit. You wanta call Dover?"

"All right." Holt started making a list on his notepad.

"And get a message to that damn pigeon that I gotta see him tonight or tomorrow night at the latest, and not in a goddam skin flick, either. Tell him it's serious, and I gotta meet where we can talk."

"Roger. Shouldn't we cover our ass a little bit and maybe get out a memo on the body-handling system, and how we gotta tighten it up?"

Good, solid Holt. Eye on the doughnut.

"You betcha, baby," said Gilmore. "There's that White House conference tomorrow. They're going to go for the head—mine. It's sure to happen. We need to shove some paper in their jaws. See if you can't make the report on Luckett sympathetic, too."

"I'll try." Holt scribbled. "What about Saigon?"

"I'll make some calls," said Gilmore. "We gotta get somebody good out there, and fast. You write out the message."

"All right."

"And all the usual. Make the normal agency checks. Christ, I hope the pigeon's going to come through on this one."

Gilmore looked wearily at Holt. The shallowness of their conversation, of their resources, lapped at his brain.

They worked through the afternoon.

Gilmore, searching for a Saigon agent already in place, made a round of calls to the FBI, Secret Service, the Bureau of Narcotics and Dangerous Drugs, all without success. He didn't blame them. Bureaucrats weighing his request against their own needs would not offer Daniel Gilmore of Task Force Washington what he needed to fill his needs in Saigon.

Customs, in part because he was still nominally a Customs agent temporarily reassigned to the task force, found a name. The Washington liaison man with the U.S. embassy in Saigon came up with an agent named Ralph Siddler.

"Let me pull his file and tell you about him," the liaison man said. "He's just recovering from malaria or something."

"Sure," Gilmore said. "Thanks a lot." Why was the s.o.b. being so cooperative?

The liaison man came back to the phone and said, "Ah, oh my yes, Gilmore, old buddy, you're going to like this fella right much. Right much."

"Okay you bastard, spill it."

"Lessee, he's fifty-two it looks like, been in the service since after the war—that's War Two, not War One. He lives in Saigon. Been his permanent address for most of a decade. Just a good old boy, the way they used to make Customs agents. That's about all I got here on paper."

"Thanks," Gilmore said. He hung up. Turning to Holt, he said, "They've got us some guy named Siddler, or Twiddler, some retread. He'll get us just exactly zip point zero."

"Maybe. You don't know with guys like that."

"You know him."

"Yep. All us prewar types know guys like that," Holt said. "We drifted into this business more or less by accident, and in those days it was steady work, simple work. Maybe he ran some guns or dealt some dope way back then. Yeah, I know him. Used to drink vodka before it got fashionable. Back in the fifties he got in some scrape with a bureaucrat back here—hadn't accounted for some expense money or was rude to a junketing congressman. The service was smaller then. It was the kind of thing drives them wild back here, doesn't mean much out there. I think he weathered it, but it pretty much ended his career. I mean, any thought he'd get ahead. He's a good old Customs man with broken arches."

"Well, goddammit, Bob, we don't have to put up with that," Gilmore said. "I can tell the Customs rep at the morning meeting this guy is unacceptable."

"There's one other thing about him you oughta know."

"What's that?"

Holt became fatherly. "He's a real sombitch, Dan. They may kill a thousand kids like Luckett, but they ain't no way they're going to kill that fat bastard Siddler. He's been around too long and his hide's too tough. He's survived a damn long time out there. That's *some* kind of plus. And there's another. Guys like that have a way of picking up information that new guys will take months to learn. Maybe the new guys never learn it. The old guys pick it up over bent elbows in bars and bent knees in whorehouses. It's a craft that takes a lifetime to learn."

"Okay, I get the point. No more bright young men for a while."

Over the next half hour they finished the cable for Siddler:

URGENT//SECRET AND CONFIDENTIAL
PRO CUSTOMS RALPH SIDDLER
DE CUSTOMS CONUS//DANIEL GILMORE TASK FORCE
WASHINGTON
U.S. EMBASSY SAIGON
INFO ADDEE: COORDINATING COMMITTEE FOR TFW,
WHITE HOUSE

MOST URGENT. REQUEST MOST URGENT INQUIRY INTO CIR-
CUMSTANCES OF DISAPPEARANCE AND DEATH OF CUSTOMS
AGENT CHARLES ARTHUR LUCKETT, SOCSEC NO. ONE EIGHT
NINE TWO THREE SEVEN SEVEN SEVEN NINE, ON OR ABOUT
ONE JUNE TO FIVE JUNE. LUCKETT SENT TO SAIGON CO-
VERTLY FIFTEEN MAY TO INVESTIGATE SOURCE OF TYPHOON
BRAND HEROIN NOW REACHING FEDERAL CITY. LAST MES-
SAGE FROM LUCKETT ONE JUNE. LUCKETT BODY FOUND
THIS DAY IN COFFIN IDENTIFIED AS THAT OF PVT JOSEPH
DIMALCO, SOCSEC NO. FIVE TWO SIX SIX SIX ZERO FIVE NINE
ONE. AUTOPSY CAUSE OF DEATH STAB WOUND IN THROAT
BY PRO. BODY DISFIGURED TO HIDE IDENT. LOOK INTO TAN-
SONNHUT BASE MORGUE AS PART OF INQUIRY. PRESUME
BODY MISTAKENLY IDENT THERE. LUCKETT CASE HAS HIGH-
EST GOVERNMENT INTEREST. REPEAT HIGHEST. ADDITIONAL
DETAILS EN ROUTE VIA SCHEDULED COURIER POUCH NO.

FOUR FIVE SIX TEE BEE TWO. IMMEDIATE ACTION DEMAND. REPEAT IMMEDIATE. DANIEL GILMORE, ACTING DIRECTOR TASK FORCE WASHINGTON.

"That oughta get him off his ass," said Gilmore.

"Yeah, if he can understand it." Holt took the telegram and went down the hall to get a courier to run it over to the State Department for encoding and transmission.

When Holt got back, Gilmore was slumped in a corner armchair, a dilapidated affair with stuffing falling out of one of the arms.

"There's a message that the pigeon will see you tomorrow night at eight in Tony's Guardhouse," said Holt.

"Okay."

"There's one other thing, Dan," said Holt.

Gilmore didn't reply. At length, he said, "Yeah, I know. Valerie Luckett."

"You want me to do it?"

"No, I'll do it. It's the least I could do, isn't it?"

2

Valerie Luckett lived at a Foggy Bottom address. Gilmore walked there, seeking time to think, going slowly through the thinning homeward-bound crowds, watching the people in their elaborate attempts not to see each other as they waited silent and polite at the bus stops.

He tried to shape the words he needed to tell a wife that her husband was dead—that it had been an event of mortal struggle, then terror, unattended by the Lord and the Heavenly Host. It wasn't supposed to be that way when you were twenty-five and newly married.

The city was emptying, slowing, quieting. The walking was soothing him. He reached the townhouse on Eye Street where Luckett had lived. Four pots of geraniums were set out on the cast-iron steps. A new MGB in British racing green and with freshly minted District of Columbia plates stood at the curb. It all spoke of tidy, soaring hopes, just like Luckett with his ambitious confidence in himself as the successful team athlete, the positive gentleman.

Gilmore went up the steps and rang the doorbell. The sky was turning soft blue and pink as the sun went under. He waited, and after a while he could hear a light tread within.

"Who is it?" the voice from the telephone asked.

"Dan Gilmore from the task force."

There was a pause, then the door opened. She was tall, dressed in a short white skirt, light yellow jersey halter, and tennis shoes. Reddish blond hair, seagray eyes that searched his face very quickly and seemed to guess why he was there.

"I thought you would be much older," she said and led the way inside, subdued and distracted.

He followed, watching the sway of the skirt against the long, tanned legs and the way she turned, with a quick move of narrow hips.

Gilmore's guess had been right. Charley had got himself a good woman, and Gilmore found himself wondering at the skewed vectors of his own life that had kept him from doing as well.

In the living room she turned and faced him, jarring him back to his purpose.

"Please sit down, Mr. Gilmore." He remained standing.

"Charley's dead, Mrs. Luckett."

She sat on her beige modern sofa, sat there on the edge, her hands folded.

"Please tell me about it." Her voice had a calm, business-like tone, and Gilmore remembered how the news of death often worked on you, how you knew the truth the instant before the words were spoken, how everything stiffened inside you and nothing seemed real, how you kept talking and acting as if discussing some real estate transaction, and how, later, alone in the dark perhaps, when you finally had to face it directly, the sobbing and hysteria rolled in and carried you away.

"Yes," he said, "I'll tell you what I can, but you must not repeat it to anyone."

"Yes," she said.

"Charley died in Vietnam. He was on the trail of a major narcotics ring. The men he was trailing killed him. He died instantly."

"I see," she said, again motioning him to sit down. He did. "Where is he now?"

"At Walter Reed. I'm . . . I'm sorry, Mrs. Luckett."

"It's not your fault," she said. "Can I see him?"

"Yes, he's . . . he's . . . ready for viewing." Gilmore couldn't bring himself to tell her about the disfigurement. Let them do that up at Walter Reed, let them tell her they couldn't lift the towel, that she couldn't see his face ever again.

She wanted to hear more, and he told her what he could: that Charley had showed up in a coffin, that they had learned of it from an informer inside a drug ring run by a man called Nick, that they were working on Nick and had another agent working in Vietnam, but that Charley's death was still a mystery wrapped in the larger mystery of increasing heroin flows into Washington from Southeast Asia.

"You and Charley have been working on that for a long time." She almost seemed to be trying to comfort him.

"Yes, Mrs. Luckett, we have."

He turned to routine matters. Did she have enough money right now? Would she be available next week for a clerk to come and explain her pension and other rights? Yes to both of these—but no, she didn't need a female agent to come and stay with her for the night. It was kind, but she was all right. She wanted to be alone.

He rose to leave, but she returned to the circumstances of the death, dissatisfied. He repeated, then told her he would call her and tell her as soon as they learned more. He placed his card on the coffee table beside her.

"You know," she said, "I have been . . . expecting this."

"Yes."

"You all live—you and Charley—by violence. It could have happened . . . at any time."

She looked up at him, and neither spoke.

Then she said, "Thank you, Mr. Gilmore."

Gilmore walked to the door and let himself out into the gloom, feeling sour, exhausted, and, in some ancient way, unworthy.

Gilmore hailed a cab and went home to Capitol Hill, to his small apartment in a once elegant building several blocks down East Capitol Street from the Supreme Court. The apartment was two flights up, convenient to a neighborhood market run by a family of Chaldean thieves. Two big air conditioners in the living room windows roared on the street side of the building. Gilmore let himself in. It was ten o'clock.

A massive stereo system and racks of records and tapes took up one wall. Across from the quad speakers was a black leather Eames lounger. Anita sat there, a black-haired six-footer with the powerful legs, torso, and upper body of a professional dancer. She was sipping a gin and tonic and reading *Newsweek*. Her summer leotard top pressed against her body, a smooth elastic glove that revealed the rich curves of her breasts, the convex span from rib cage to broad dancer hips, and the secret swellings below the curve of her belly. The smell of pot hung in the air. She looked amiably at him and waved.

Oh yes, my other self, my goaty me, my need, my hell, my ecstasy. Gilmore pulled off his coat and slung it on a chair, took the .38 Smith and Wesson Police Special from its clip holster at the base of his spine and laid it on the coffee table. He loosened his tie and went into the kitchen where he rustled a deep fist of tangy dark rum and a head of tonic over as many ice cubes as he could cram into one of the muglike glasses he preferred.

He swirled it and drank it, feeling the cool liquid sizzling down, pumping energy. His stomach growled. He opened a can of jellied consommé, squeezed a slice of lemon over it, and doused it with Worcestershire sauce, then mixed himself another drink. The stereo went on while he was adding tonic, the pleasant arabesques of the Modern Jazz Quartet. He went back into the living room and sat in the lounger. Nita was doing her exercises, the slow ones that meant she would want him to make love soon, so they could make love again soon after that.

Gilmore smiled, shaking his head. "Jesus, my lovely Nita, let me eat my consommé."

"Sure," she said, her nose to the floor, legs spread wide in a graceful deep curtsy. He slowly downed the soup and drank the rum, watching her go through the limbering techniques. Despite her size—she was much bigger than most dancers—she had nearly perfect control of her limbs. Gilmore felt his tension subside as her shapes and forms unfolded before him. He finished the drink and lay back against the cool leather of the lounger, even forgetting the White House meeting he faced in the morning.

In a few more moments, Nita moved over to him and unbuttoned his shirt, her long fingernails like icy points as she ran them across his chest. The fingers stopped at a ragged scar over his left collarbone where a drug pusher named Roland Drinkwater had carved with his switchblade.

Gilmore had killed Roland Drinkwater, the slaying that had brought Gilmore both headlines and the skin flick pigeon. When Nita stroked the scar, the scene came alive again: Drinkwater's back as he fled down a darkened alley five blocks from Gilmore's apartment; the chase; struggling; and then the sharp pain where Drinkwater pushed in the switchblade and the gritty sting when the blade carved into the collarbone. Gilmore was down on his back and somehow the gun was in his right hand. He raised it and fired, the muzzle flaring orange in the winter gloom. The flat-nosed, high-velocity mushrooming slug had gone in at the rear of Drinkwater's right shoulder and split into two pieces on the bone. One piece gouged down under the arm, shredding out the triceps muscle. The other piece deflected into the rib cage, surgery later showed, and tumbled through the upper lobe of the right lung, coming to rest against the heart.

Drinkwater had died three days later in the intensive care unit of D.C. General. Gilmore had not wanted to kill Roland Drinkwater. Gilmore did not like the high-velocity slugs. The shooting had made brief headlines, and shortly thereafter, a man had telephoned the task force number, asking for Gilmore. The man was an informant—the skin flick pigeon.

"You're my love, my own," Nita breathed, her voice soft and humid in his ear. Gilmore reached up, stripped the leotard top down, and watched the taut planes of muscle gliding beneath her skin as she pushed it the rest of the way down and stepped out, showing herself to him.

"You are incredible," he said.

She bent over, her breasts swinging, and took his face in her hands and kissed him. Her mouth was warm.

He kissed her long and hard. They headed slowly into the bedroom, pausing occasionally while he took his clothing off.

My darling, he thought, you're no Valerie Luckett, but you do me nicely.

Gilmore gave himself up to her as she flexed and expended herself. She drowsed against him, then stirred and sought more love. He took the lead, making her shudder in an excitement that oiled their bodies with sweat and lifted them toward a brilliant spasm of energy followed by almost instant sleep.

3

Gilmore settled into his chair at the conference table at seven o'clock the next morning. His mouth was dry and his temples throbbed. He felt hung over, depressed, and angry. He and five other men had come together in the Executive Office Building next to the White House for the task force's weekly meeting.

John Hathaway, an inspector of the Washington metropolitan police, was sitting opposite Gilmore. Hathaway wore reading glasses and thumbed through the pages of Gilmore's report on Luckett with a rawboned thumb the size of a spaniel's foot. The other four were still reading when Hathaway looked up and took his glasses off, revealing his large beefsteak face in all its glory.

"So he got killed in Vietnam," Hathaway said. "You sent your buddy out there and he got himself killed. That's too goddam bad, Danny fella."

"The coordinating committee approved sending an agent there," Gilmore said wearily.

"Yeah, but *we* didn't choose him—*you* did. Right?"

"That's right. I did."

"Just so we keep it straight, Danny." Hathaway sat back and sucked at his coffee through pursed lips. The sound seemed to stir the committee chairman, Noel Walker, a special assistant to the President for law enforcement and internal security. He looked around in annoyance and riveted Hathaway with his glance.

Walker's thin, sun-seamed face was as dry as the southwestern plains from which he came. He was a narrow, weedy man with mousy hair and eyes faded from years of farming and politicking across his state. A drought at the polls two years before had squeezed any humor from him. He presided with a thin, parched voice which he now used. "Who's next?" It was said with an inflection that withered volunteers.

"I got a question." Vincent Axby of the Bureau of Narcotics and Dangerous Drugs, a sallow-faced, professional infighter, looked up. He didn't wait for Walker to nod assent. "This report says there's a man in Saigon looking into Luckett's death. Now who would that be? You didn't send another man from here, did you?"

"No." Gilmore wanted to punch Axby and bring some color to his face.

"You learn from your mistakes in other words," Axby said. "That's good. That'll look good on your fitness report, Danny, or maybe you young geniuses don't have fitness reports the way the rest of us do. Who's the man in Saigon, Danny?"

"A guy called Siddler. Ralph Siddler."

"Siddler!" Axby guffawed. "Syphilitic Siddler? Are you sure, Danny?"

Gilmore felt his weariness turn to fury. These were the jackals who had sensed the weakness in Freddie Moran and pulled him down. They had sent Moran to a place where people dry you out and get some strength back in your legs and some sawdust back in your head. But they never can repair the broken part that stands at the center of your being and gives it whatever poise and strength and grace you have. There's no repairing that part once it's broken.

Gilmore glanced at Walker. The faded eyes were shrewd and expressionless. Plains lizard. Walker had watched Moran brought down. If Gilmore were pulled down, he would be replaced as Moran had been. Walker was looking somewhere else. Maybe he wanted to be a congressman again, or perhaps a senator. He would run a strong law-and-order campaign. He was not interested in Daniel Gilmore. He paid no attention to Axby, but sat looking at an open loose-leaf notebook. There was silence while the others waited for him to talk.

At last he said, "What are the current estimates on drug addicts by the end of the year?"

It was an irrelevant question. Why doesn't the s.o.b. say something about Luckett?

Hathaway rummaged in his papers. "Metro intelligence puts it about one hundred new addicts a month, or twelve to fifteen hundred by the end of the year."

"What's the estimated city total for the year end?"

"Maybe fourteen thousand."

"Last year's total?"

"About the same."

Walker paused and looked at each of them—Gilmore, Axby, Hathaway, and two others who had not spoken, Albert Flann, a wispy white senior bureaucrat from Customs, and Henry Elliott, a black lawyer from Justice.

"There's been no improvement then, has there?"

"The year isn't over yet," Hathaway replied.

"You doubt your own estimates?"

"Those are only estimates . . . they can change."

Walker turned to Axby. "What does BNDD estimate?"

"Twenty thousand."

"For Washington?"

"Yessir," Axby said. He was regarding Walker closely,

apparently deciding which way to jump on the issue. Gilmore watched him do a bureaucratic squirm. Then Axby said, "That's the low end of the figure, sir. Other bureau analysts feel it's up as high as twenty-five thousand. For the city alone."

"What do you say, Gilmore?"

"Our intelligence comes chiefly from metro—"

"What do you *say?* I don't want a lot of preface."

"We think it's the lower number."

Walker studied his notebook some more. Then he said, "Who knows about the Luckett slaying?"

"His wife and the task force," Gilmore said.

"What about the family?"

"He didn't have any to speak of. A sister on the West Coast. We haven't found her yet. She's married, or remarried. His dependent sheet listed only his wife."

"Anyone else know of this?"

"The Saigon man—Siddler. And the family of the other boy at the funeral yesterday. They know something happened, that the body wasn't their son."

"But you've shut them up?" It was a statement rather than a question.

"They've been told that it's a matter of highest national security. They know they're not to talk about what they saw at the cemetery."

"That gives you how much time before the media gets the story?"

"I'm not sure I understand the question," Gilmore said. Where was Walker heading?

"How long? How long?" Walker glared. "That's a simple question, goddam it. How long before they go to the papers with the story?"

"I don't think they will," Gilmore said.

"Oh, good Christ!" Walker said in exasperation. "Don't

you know anything about human nature? Of course they will. There's been no funeral, boy. *No funeral.* They wanted to cry and now they're confused. They want some action. They'll go to the papers, that's what they'll do. A cover story should be prepared. Make sure the Pentagon knows and has something ready to go. They're good at that sort of thing."

"All right."

Walker looked again at the notebook. "Are there any leads on the Typhoon heroin?" The question was greeted with silence. "Don't all speak at once. Any leads?"

"We may have one," Gilmore said.

"What do you mean by that?" Walker demanded. "Is there something I can tell the President if he wants a briefing on this mess?"

"We're close to getting Nick," said Gilmore, his jaws tight with the effort of prevarication. "The pigeon's information is getting better. He's setting up a buy."

"Oh bullshit," said Walker. "The information you've been giving us from that man hasn't been worth a hot flying damn."

"He told us about Luckett," said Gilmore.

"Nick killed Luckett I suppose," said Walker.

"His people in Asia did. The point is, Nick is operating in this Typhoon stuff—we all know that. The pigeon says a big connection is coming up, and he's setting us up for a buy."

"We know about that shipment," Axby said. The sallow face was shadowed with doubt. Gilmore knew it was a bluff. "We're working on it. There's no need for the task force to work it, too."

"I must say," Flann put in, "Customs knows nothing of this so-called major connection. Nothing."

Walker turned to Elliott, the Justice Department lawyer,

and said with a sardonic smile, "What about you, Henry? You got some ideas on the subject?"

"If these guys disagree, then I don't know what to believe," he answered.

Walker laced his hands in front of him. Gilmore saw that his dusty face was turning pink with anger. "It's time for a summary," Walker said. "This is the thirty-fourth weekly meeting of this group. There have always been troubles among the participating members, but in the past, these troubles were the usual nonsense we have come to expect from entrenched bureaucrats. To some extent, it's understandable. You've each given up several valuable people to this effort and then had to face the fact that the task force was making headlines and arrests on its own, more or less in competition with your own agencies.

"As the pressure has come down in the federal city, as the eyes of the nation turned to the spectacle of a new wave of heroin pouring in, your jealousies have become sharper.

"This happens to be an off-year election for an administration that has brought millions of dollars to you damned bureaucrats in the fight on drugs. We've got the public behind us, the Congress, the courts. And you know what? You bastards spend half the money and half the time fighting each other.

"Look at this Luckett thing. That's a goddam dangerous situation to this administration. The press is going to get hold of it, and what are you guys doing? Trying to hang it on Gilmore.

"Well, any dummy can see he doesn't need it hung on him—by Jesus, it's there." His voice blew thinner. "Here's some more stuff you call facts: addicts in Washington, anywhere from fourteen to twenty-five thousand, depending on which liar you believe. No leads on this Typhoon heroin.

Gilmore scraping around with some story about an informer who's going to unlock the whole mess.

"Now this stuff is going to cease. It's finished. You understand me? This man Siddler, I never heard of him but I got my doubts, too. So I sit here and listen to you and I ask myself: why didn't they get Gilmore somebody good, some crackerjack fella that's going to clean this thing up?"

Walker's eyes were reddening with the effort; his face glowed a subtle pink. "Now we're going to get down to business. I'm giving you one week, Gilmore, to make some progress on this Luckett case. One week. We're going to tell our President in one week that we're getting close on this thing, that's what we're going to do. You get that?"

"Yes." Gilmore's innards knotted.

"You goddam well better, because if you fail me, I'll bust you right down to your rich little shoes." He turned his glare toward Axby and Hathaway, catching them in mid-smirk.

"Now you two, you're like old dinosaurs—dangerous as hell but goddam near extinct. I'm here to tell you to start doing your jobs better or you'll be gone, that's all—with nothing but a few old footprints and some bones left behind. Now that's all I got to say. That's the end of it. We meet in a week and there'll be results or out you go. We're adjourned." Walker got up and stalked out, followed by Flann and Elliott.

Axby and Hathaway smiled easily at Gilmore. "Jesus, he's a mean one," said Axby. "I didn't know he had such a temper. Gets riled, doesn't he?"

4

That evening, when the summer air was bluegray with oil haze from the rush hour traffic and the sun was brassy and wild as it fell, Gilmore set out on foot for Tony's Guardhouse. He was angry; he knew he had wasted the day waiting for his meeting with the pigeon.

He walked up to Thomas Circle at Sixteenth and Massachusetts Avenue and glanced south. The White House lay six blocks away, low-slung against the smoggy sky, almost hidden behind a high, wrought-iron fence. The flag hung limply from the North Portico. There, the whites in the country had access to the boardrooms and conference halls and treaty chambers that ratified and legalized corporate plunder, the places where public policy and private inter- est merged out of sight of prying eyes.

Across Thomas Circle to the north, lay another human jungle—this one gasping in overheated neighborhoods that stretched for nearly ten miles to the north and east. There, the blacks had access to the streets, and their plunder was as direct and comprehensible as a knife or a bullet. It made easy newspaper copy and it preoccupied the minds of the whites, stalled in traffic jams, as they stared moodily at the buildings lining the major arteries. Presidential power— power that could order the building of dams, the leveling of foreign cities, the virtual remaking of huge portions of the globe—had little impact on the street ten blocks from the White House.

On those hidden streets, it was the hour for the stirrings of the murphy men, the pimps and their prostitutes, the pushers and the yoke men. The man with the golden spoon was snorting coke in his lair, working himself to a fever pitch to light his way through the night to come.

Tony's Guardhouse was in the basement of a new high-rise office building on Massachusetts Avenue. It attracted hard, bright couples of various races. The entrance from the street was a glorified manhole, twelve feet down by way of circular iron steps.

Gilmore went down and pulled open the door, an oak front affair that fit like the door of a safe. He could hear the driving, erotic beat of hard rock.

The Guardhouse was a long room lit with dim lightbulbs set in chrome ceiling-clusters against metal foil panels. The walls were dark-rimmed mirrors, the floor black, the booths around the edge of the place black vinyl, chrome, and dark formica. The bar was a wide, plush U-shape of polished fruitwood in the middle of the room.

Gilmore ordered a rum and tonic from a female bartender with a blouse split down to her navel. When the drink came he went over to the last booth in the corner, slid in, and settled against the wall. The fire exit was between him and the bar, set between the bathroom doors in a back wall. The place was half-full with patrons who had escaped from their hard-edged worlds and liked drinking in a hard-edged place.

He was on his second rum and tonic when the pigeon suddenly materialized at the other end of the room. Gilmore checked his watch. Eight o'clock.

The pigeon—a man with a pointed face and small goatee
—was wearing soft purple bell bottoms and an off-white,
double-knit jacket over a dark blue shirt. He was sweating heavily, the moisture soaking through his shirt front.
He slid quickly into the booth across from Gilmore.

"I'm in trouble," he said in his hoarse whisper.

"What do you mean?"

"Nick suspects something. We gotta cool it. I think he
may even have some dudes following me."

Gilmore felt his gut tighten and flutter. This was no time
to be cooling it, no time to be backing off. He leaned forward toward the sweating face.

"Now relax," he said. "Just relax. You're all right. We'll
take care of you. I just need a little information—like how
did you find out about the DiMalco coffin?"

"A tip."

"What kind of tip?"

"Something Nick said. I don't remember exactly—but
someone in Vietnam told him, something about a body
getting loose that might blow their operation."

"Nothing more solid than that?"

"No, just the DiMalco name."

"You don't know who killed Luckett?"

"Who's Luckett?"

A cocktail waitress came by but Gilmore waved her
away.

"Jesus, man, I told you, I can't—"

And then Gilmore saw them, three blacks coming
quickly through the front door, dressed in dark clothes,
dark shirts, maybe turtlenecks, it was hard to see in the dim
light. It was the gloves that tipped them. Leather gloves.
They stood for a moment, thirty feet away, scanning the
place as he had done earlier. Hunters.

"HANDS UP!" Gilmore shouted as the first man pulled an Army .45 semiautomatic from his coat pocket and started toward them.

Gilmore rolled off his own pistol, fumbled for it, moving across the booth. The .45 was trained in their direction, and the report thundered, going high; maybe the hunter was a little nervous with so many people around. The slug shattered the mirror behind them. People were shrieking and scrambling to get down. Gilmore's .38 was free. He fired and the .45 answered and then answered again. The last slug hit the informant in the abdomen. The man screamed and bucked in agony.

Gilmore rolled out onto the floor, firing wildly twice. Get their goddam heads down, get some time. The informant was probably dying; Gilmore could hear him struggling with his wound. Gilmore scuttled for the fire exit. Another man came toward him. Another .45. Gilmore shot him in the chest. The bullet made a heavy slapping noise when it hit the breastbone. The man went over backward and his semiautomatic spun in an arc to the floor. Like a rock, Gilmore thought. I saw it turn in the air. Down like a rock.

A gunman suddenly appeared around the end of the bar, prone, sprawling toward Gilmore, his arm stiff in front. The distance was ten feet. The .45 roared, flashing in Gilmore's face. Jesus Christ save me. The doe-eyed bartender stuck her head through the narrow entrance behind the bar and screamed. She had been shot in the face. She pitched forward. The gunman seemed startled. He hesitated, perhaps blinded by the muzzle flash of his own gun.

Gilmore fired once. The bullet went into a booth across the room. There was a commotion over there. He fired again. The second bullet hit the gunman in the forehead.

He went down without a sound, twitched once, and lay still.

Gilmore jammed himself against the exit door. It eased open and he backed into the stairwell. A .45 boomed from the opposite end of the bar. The slug hit the jamb in front of his face, showering splinters into his face.

"Jesus! I can't see!" He fell heavily against the first steps of the fire stairs, then groped his way up through a reddish haze. He felt frantically for the outside exit door, and at first there seemed to be no way out. Then, squinting, he saw the door right in front of him. He pushed it open and, as he dived through to the alley, heard the gunman pounding up the fire stairs screaming at him.

Gilmore stumbled out into a darkened alley between the office building and a small professional building. He ran along it, groping to reload the police special.

He fell over a row of garbage cans and went down with a clatter, the cans and lids rolling drunkenly. He started up. The .45 boomed from the fire exit doorway. The slug ripped a fist-sized hole through a trash can and splatted into the brick wall.

Gilmore scrambled up, tripped over a can lid, and went down on one knee as the .45 fired again and the slug went where his head had been. He lunged up and began running up the alley, running toward the street light at the end. He tried to weave and keep low, but seemed to remain in slow motion, a treadmill. Legs in mud. The lead would be fire hot, cutting through the back muscles, into viscera. The street light cast a long shadow.

A figure half-appeared. The shadow was long, stretching down the alley—bisected torso, one arm, one leg, half a head. The arm was pointed forward, the figure clung to the wall. Gilmore flinched as the pistol in the man's hand snapped three times quickly, the muzzle flaring bright against the wall.

"Gilmore!"

"Watch out! He has a gun!"

"I got him! I got the son of a bitch! I heard it hit!"

"Holt! Jesus Christ! I thought I was dead!"

"You stupid shit, you oughtta be!" Holt crouched over him, peering into the darkness. They could hear someone breathing heavily, groaning. "A lung shot," Holt said. "You can hear him wheezing."

"You weren't supposed to come here."

Holt shrugged. "I wandered by on my way home."

A siren sounded far away, then another and another, growing louder.

"Stay out here and tell them I'm not a crook," Gilmore said. "I'll go see if that guy can talk. His buddies aren't going to."

"What happened in there?"

"Three guys came in. They shot the pigeon."

"Bad luck," Holt said. "Bad goddam luck."

Crouching, Gilmore moved quickly down the alley, holding the reloaded pistol in front of him. He knelt over the sprawled man who was lying face up under the exit door. The man's lips moved in a slack face. Gilmore listened. ". . . Stupid . . . stupid . . ." the man said and coughed once.

"Bring a box," Gilmore said, getting up. "This one's dead."

The alley filled quickly with police and rescue squad attendants. A black lieutenant was in charge. He scrutinized Gilmore's ID and then said, "How many dead?"

"Three more, I think."

"Jesus," said the lieutenant. "What'd you use, a machine gun?"

"One's theirs. Two're mine. One's my partner's."

They went through the pockets of the dead gunman in the alley and found nothing. An identification team rolled a set of fingerprints and took Polaroid photographs of the body, then trotted toward a mobile crime lab van parked out front. "We should know who he is in a few minutes," the lieutenant said. "He looks like a man who ought to be on file somewhere."

They went downstairs. Gilmore wet his handkerchief and wiped the blood from his face and eyelids. The Guardhouse was in an uproar as terrified patrons watched police and rescue squad workers ministering to the wounded.

The bartender was covered with a coat and there was an oxygen mask over her face. She was in shock, breathing fast. The attendants had an intravenous tube in her and were getting a stretcher ready. A man lay on one of the tables, moaning, while an attendant worked over his knee. It had been shattered by the stray bullet from Gilmore's gun. Several police were interviewing other patrons.

They turned to the dead. The informant lay half under the booth, face down. The back of his jacket was mottled with a wide red stain. Gilmore fished the man's wallet from his coat, flipped it open, and spread the contents on the table. There were two twenty-dollar bills, some singles, and a Texaco credit card. An identification card gave the man's name as Guy Burton, with the address in the Columbia Road area.

"Who's this one, Max?" the lieutenant said to a uniformed cop who was kneeling over the body of the gunman who had been shot in the head.

"I don't know. Radio man is checking his ID," Max

said. "But that kid over there, the one with the bullet in the chest, that's a kid named Peter Parkes. I'd know him anywhere—dead or alive."

"Why?" the lieutenant asked.

"Because he used to be a magician and give shows, you know, for kids. He's known as 'Roulette' Parkes now, a big gambler."

Gilmore and Holt looked at the kid—he was lying on his back, arms and legs akimbo. He looked to be eighteen or so and his face in death was serene.

"Where's he live?" Holt said.

"Used to live over in Anacostia," Max said. "Then he moved into the Columbia Road area about a year or two ago, struck out on his own. He's been arrested two, three times. The reason I know about him is his mother called me a few times after his first arrest. I went over to see him. It didn't do no good. He was a tough little shit, that's what he turned into."

A uniformed man came in the door and went to the lieutenant.

"Man in the alley is Raymond Castle," he said. "Ident office says he's known as Tiger. Tiger Castle. He's twenty-five, a gambler, a hustler, and lately into dope. Married at age sixteen, had a juvenile record since he was thirteen, convicted at nineteen of assault and served eighteen months. He's said to be an enforcer, possibly for Big Nick Westley.

"The other guy there is Louis Michael, nickname Perez, because he's half-Chicano. Former GI from San Fran, busted twice, went AWOL, and was wanted for shooting a private security guard in a break in at a National Guard armory on the Eastern Shore. He was also known at one time to be working for Nick."

"Okay," said the lieutenant, turning to Gilmore. "Now what in hell happened here?"

Gilmore sat down in a booth and sketched in his meeting with the informant and the arrival of the gunmen. As he talked, his spirits dropped. The informant was dead; there would be no deal with Big Nick.

The lieutenant took notes and Gilmore listened to his own words and watched the cop write things down. His mind wandered back over the White House meeting and settled bitterly on Axby and Hathaway and he heard again the unpleasant whine of Noel Walker.

Events were trapping him, he wasn't getting any breaks. Holt hung back in the gloom talking to Max quietly, apparently unconcerned.

As Gilmore finished, the attendants removed the two wounded people and the bodies were shrouded and carried out. The press was gathered at the front door, arguing with a policeman there. The lieutenant closed his notebook. He was unhappy; Gilmore had refused to tell him why he was drinking with Guy Burton. The officer wandered back to the kitchen and called the station house.

Holt slid into the booth. "Dan, let's get out of here. Now. I got the address of this Roulette Parkes out of Maxie over there. We're going to pay a call on Roulette's old lady—if he has one."

Gilmore looked at him dully. "So what?"

"What do you mean, so what?" Holt said. "It's the only lead we've got. We're going up there . . . we'll beat the cops by half an hour if we move now. They gotta get a legal search warrant. We can be gone by the time they arrive."

5

They took Holt's personal car, a battered Chevelle sedan, and in a few minutes had raced through sparse nighttime traffic to a small sidestreet off Columbia Road. Roulette Parkes' address was an apartment house that stood in the middle of the block, the only inhabited structure in the area. It was dilapidated brick: broken windows behind battered heavy mesh screens on the first floor, a weed-choked front lawn glinting with glass and flowered with newspapers and garbage.

They circled the building, then parked a half-block away off the main street. They walked back slowly and went up to the front entrance. The double doors into the building hung askew, broken at the hinges. A tattered old man sat on the wide concrete front stoop, staring vacantly at them. An empty pint of Park and Tilford Reserve lay next to him. He muttered to himself, unseeing.

They went inside. The lobby was strewn with newspapers. Under the rifled postal boxes lay a thin black man, sweating and shivering, his eyes slits, his head moving slightly. "Scag," Gilmore said.

They went upstairs through a stairwell loaded with trash and rotting garbage and bitter with the stench of urine.

"What's a fancy gambler doing in a place like this?" Gilmore asked.

"Beats me. Maybe it's the wrong address."

The sounds of rock boomed down the empty corridor on the second floor. Three dim lights in wire cages lit the hall. Spray-can graffiti littered the walls with obscenities. They moved toward apartment 9, the address given for Roulette Parkes. Gilmore pulled his pistol.

The door to apartment 7 opened slowly and a young, odd-looking boy walked out. He froze, his eyes bugging at the two Customs agents. "JESUS!" he screamed, and fell on the floor, writhing. He was panting and gasping, his eyes rolled back in his head, teeth clenched, jaws bulging, his body wracked with convulsions.

"Christ, he's having a fit," Gilmore said. The door popped open again and a seamed face gave them a malevolent look. A fat arm shot out, grabbed the kid by the foot, and dragged him inside. The door slammed and locks were thrust home.

They went down to apartment 9. The music that filled the hall pulsed from behind its door. Gilmore banged his pistol butt on the cheap door. "Open up! It's the police! Open up!" He waited a moment, then stepped back and kicked the door. It flew open and the smell of pot drifted out.

The room was dark. Gilmore reached around the jamb, found the light switch, and flipped it. They barged in. Two girls in halters and jeans sat on a broken couch against one wall. There was a boy sitting between them. He was stark naked and had a roach in his mouth.

"Shee-it," he said.

"Up! Up!" Holt ordered. They yanked the three to their feet and quickly patted them down, searching for the touch of a knife handle or Saturday night special, feeling the girls' breasts and thighs, pressing their hands between their legs. The sullen girls said nothing. The kid's face was expressionless.

"Peter Parkes live here?" Holt demanded. The kid said
nothing. Holt slapped him full in the face. The kid's eyes
reddened and tears started. Holt slapped him again. The
youth's head went down and came up.

"Who're you?"

"Brother."

"Roulette's brother? What's your name?"

"Paul."

"Okay, brother Paul," Holt said. "You're in deep trouble.
Deep goddam trouble."

The customs men sat the three kids on the floor, their
backs to each other, and handcuffed them together. Then
they ransacked the apartment.

The apartment consisted of a small living room, a cell-
like bedroom, kitchenette, and bath. It was furnished with
the broken couch, a scarred bureau, a card table, four tubu-
lar chairs, a small coffee table, an expensive color television
set, and a large stereo rig surrounded by stacks of records
and tape cassettes.

The bedroom contained a steel-frame bed, another bureau, and a small bookcase. A small closet, stuffed with
clothing, opened off the bedroom.

The apartment's walls were decorated with circus posters
and black revolutionary art showing black men and women
wearing thick ammunition belts and little else. They were
shooting off automatic weapons.

"Looks like a college dorm," Gilmore said. They ripped
the couch apart, tearing the pillows open and turning the
frame upside down. They pried the back off the television
set, tipped the set up, and shook it, then pried the back
from the stereo and looked inside. They ripped the fronts
from the two speakers and looked there.

They pulled the drawers from the bureau and quickly

went through them. Numerous decks of cards, assorted shirts and socks and underwear. They found three newly starched and pressed U.S. Army dress shirts and a neatly pressed Army overseas cap with the insignia of the First Logistical Command on it.

In the bedroom, they pulled every piece of clothing off the hangers, went through the pockets, and threw the items in a pile on the floor. They found a green, Army private's first class dress uniform hanging in the rear of the closet. It had a bright red and blue First Logistical Command patch on the shoulder.

"Was your brother in the Army?" Holt shouted around the corner.

"Nope," said the kid.

"Maybe it was for a costume party," Gilmore said. The trousers were perfectly pressed with a knife edge crease down each pantleg.

"Boots, too," Holt said, throwing out a spit-shined pair of combat boots.

"Pass any inspection."

"Yeah. Just like they were used by a real soldier," Holt said. "Looks to me like they'd fit him, too."

"What's your brother do with these clothes?" Gilmore asked.

"Don' know," Paul Parkes replied. "Wears 'em, I guess."

"Sure, right over his pink tights," mumbled Holt.

Holt was looking into the small bookcase, taking the books out, fluttering through the pages, and throwing them in a corner. "Mostly magician's stuff . . . handbooks of magic, sleight of hand, things like that."

"It's a short step from magician to cardshark," Gilmore said.

They took the bed apart, stripping the covers, turning

the mattress, ripping it open, and looking in the ticking. They tipped the frame and worked each caster wheel off the legs. Inside the fourth leg they found a loop of string. Gilmore pulled it, tugging out a small, tightly rolled plastic envelope. The envelope contained a photograph of Roulette Parkes in his Army uniform and an official Army identification card with the same photograph sandwiched between soft clear layers of plastic.

The card identified the photograph as Paul Flynn Johnson, serial number 243–32–4235, private first class, pay grade E–3, and gave his birth date and date of expiration of enlistment.

"Very nice," Gilmore said. "It looks official as hell." He pocketed the material and they went into the kitchenette.

They looked in every pot, pulled the stove from the wall and looked behind it, pulled the drain strainer from the sink and looked there, checked the p-trap beneath to see if it had wrench marks on it, and looked in the oven and the pot drawer beneath it.

Gilmore removed the pots and then pulled up the aluminum foil liner that covered the drawer bottom. A manila envelope lay there. He removed it, opened the flap, and pulled out some papers.

"Christ, those are orders," said Holt. The orders directed Paul Flynn Johnson, private first class, First Log Command, to report for duty in Saigon not later than July 1. The orders were signed by the adjutant at Fort Myer, Virginia, across the Potomac River from Washington.

"Here's more," said Gilmore. Several other bits of paper fell from the large envelope. Three were blank. The fourth had a notation scribbled on it in pencil: "Whistle back 6–20."

"Whistle what back, for Christ's sake?" Holt asked.

"It doesn't say."

"Busy people. Back from where?"

"The base PX."

"Six-twenty. That's probably June 20."

"Next week. Something is coming back next week."

They found nothing more in the kitchenette and went into the bathroom, which they searched without success. While Holt scooped up the uniform and other military gear, Gilmore uncuffed the kids.

"Something happen to my brother?" Paul Parkes asked.

"He's just fine," Holt said.

"You guys really cops?"

"Yeah."

"What you lookin' for?"

"Just looking."

They walked out and rushed downstairs. The addict was still lying in the lobby. The drunk had gone from the front stoop.

"Let's get Army Criminal to give us an answer on how this kid came to have these orders signed all nice and proper," said Holt. "And the uniform and all the rest of it."

Gilmore considered. "You can try it," he said, "but I don't think those lazy bastards will come up with much. We better get this stuff cabled to your man Twiddler or whatever the hell his name is."

"Siddler."

"Yeah, Siddler." Gilmore corrected himself. "I hope he's got our message and is moving on it."

PART II

Siddler

6

Siddler hitched his trousers up around his fat waist. His new khakis had been falling down for a week. He squinted into the warm Saigon night as he came out of the embassy into a floodlighted compound. He lit up a Viceroy, sucked deep, and walked toward his jeep.

It's the principle of the thing, he thought. "LUCKETT BODY FOUND . . . STAB WOUND IN THROAT BY PRO . . ." By pro-what? Pro-Communist? Pro-Daughter of the American Revolution? Those smart-asses in Washington, he thought, they'll have to excuse me, but I'm particular.

An hour earlier, at midnight, Siddler had sat down at a direct-line teletype to Washington at the American embassy. "WHAT MEAN STABBED BY PRO," he had cabled Gilmore, fumbling drunkenly at the sluggish keys with his stubby, jabbing forefingers. "PRO WHAT. AM WAITING RESPONSE NOW. RGDS DIDDLER." Then he had waited, his head nodding in a self-satisfied stupor, while whoever was at the other end made what Siddler hoped would be frantic phone calls.

Siddler was a small, lost-looking figure in the clatter of the teletype room, surrounded by maps and charts and long streamers of yellow teletype paper hanging from hooks on the walls. Alone there at night, he looked like a clerk slaving in an insurance office rather than a secret agent in a war zone. Siddler was a short man with a tremendous bulge of belly, stubby legs, and short, hairy arms that swung out at an angle when he walked. His sweaty face was red and puffy from drink, his eyes beady, darting. His hair was a wisp of dirty brown over baldness that shone under the fluorescent lighting. His body heaved as he breathed thickly, giving off waves of that tremendous energy that fat men seem to radiate, the reserve energy of thousands of excess calories boiling in their bellies. Wet patches of dark brown marked the armpits of his khaki shirt.

He had noticed that he had typed "DIDDLER"—that was supposed to be "SIDDLER." No sweat. Noon in Washington. Hope Gilmore gets pulled out of some fancy lunch for this. Whoever Gilmore is.

Those goddam Washington bureaucrats hadn't even told us Luckett was in town. Didn't trust us. Now they think it's some big deal because he gets killed. How long did they expect an inexperienced kid to last in a town like this? Five minutes?

Suddenly the teletype had sprung to life again with a steady pulse, the carriage slamming to the right, circuits clicking around the world. It had tapped: "PRO MEANS PRO-FESSIONAL. GET OFF YR ASS AND GET GOING. RGDS GIL-MORE."

Still preoccupied with keeping his trousers from falling down, Siddler climbed into his battered green, topless jeep and stamped his left foot around for the start button. When he hit it the jeep belched to life and he thought, well, the mangy mother didn't blow up on me.

So what do I do now? He fumbled at the controls and the jeep lurched out into the late-night sputter of motor-bikes and tiny French cars on the broad thoroughfare in front of the embassy. Siddler traced a curving path down the street toward the bright lights and concertina wire surrounding the grounds of the South Vietnamese presidential palace. Diminutive, green-clad soldiers strolled in front of the wire, their black, toylike M-16 rifles slung over their shoulders.

Where do I start? Siddler swerved past the wire and lights and soldiers and around a corner into a darker street. "HIGHEST GOVERNMENT INTEREST." Now what is that supposed to mean? It probably means that Luckett's death is somebody's ticket to promotion back there and they need me to punch the ticket for them. Those bastards in Washington. Now their kid gets killed and they tell me. Now they need me. Well, I'll look into it tomorrow if I feel good enough. I got the whirlybeds and I ain't even in bed yet.

The next morning, Siddler rolled out of the sack at nine and drove out to the Saigon Country Club on the edge of town. To get there, he had to go through the Tansonnhut Air Base and military complex. An American MP waved him past the front gate and a little Army of the Republic of Vietnam guard standing next to the MP didn't even look at him. The ARVN was busy looking at other Vietnamese, stopping them to search their pedicabs, cars, and motorbikes for bombs and dope. Americans didn't get that sort of treatment.

Siddler pulled up in front of the club, walked through the dining room, and settled alone at a table on a shaded patio overlooking the golf course.

"Gimmie a bloody mary and peanuts," he said to the Vietnamese waiter who appeared in crisp white to take his breakfast order. Two marys and three bowls of dry, roasted peanuts were his usual.

Siddler looked out across the rolling greens. He watched

groups of fat Vietnamese businessmen teeing off, waddling down the course in their shorts and red and yellow golf shirts. They were followed by scrawny caddies who looked like war refugees.

The scene didn't surprise Siddler. He had seen it all. Through the years of anguish and sorrow that have visited this steaming land, men and women have survived and carried on their lives. Siddler knew instinctively that it took more than a few bullets flying around to keep men down. People make tender love while murder is being committed in the next room; cook and enjoy tasty meals in dirt caves with mortar rounds popping outside; delight in the antics of their children while assassins' bullets whine down the streets nearby; and make financial investments in a future governed by random violence and overseen by no man's gods, a future promising only extinction.

Siddler threw his head back and snarfed in a dozen peanuts. Yeah, he'd seen it all. Slow-moving peasants serenely herded their oxen home through watery rice paddies at dusk, oblivious to the chatter of machine guns in nearby palm groves; rockets fell on cities: who would die? Men were beasts one moment and heroes the next. Poets became rapists. Men spewed rockets and napalm on the land after marveling at the beauty of what they were about to destroy.

So? The waiter set another mary down. Siddler swigged it. So what? So Ralph Siddler, special agent with the U.S. Customs Service, a man who had been idealistic in his youth, breakfasted on alcohol and nuts and contemplated what he would do about a dead guy called Luckett. What a name.

Siddler thought about Gilmore's cable. It was a cable the demands of which, Siddler sensed with a weary certainty, could only be met by the full exercise of his well-developed and brutally direct instincts. It was going to be a long drudge of a day.

His career had been twenty-five years of long, drudgy days. Perhaps he had been mildly successful. Would have been very successful, in fact, if he'd been able to survive the inner-agency politics in Washington and elsewhere. He hadn't. A jealous colleague had reported a small, damaging fact. Syphilis. They had cured it in a week but it had ruined his career. He had never forgotten that colonel's daughter from Poughkeepsie. Where is she now? he sometimes wondered. Probably in Hackensack. Siddler had always had to take it where he could get it, while his slick comrades were out screwing the dollies of their choice. Gilmore is probably one of those sombitches, he thought. Think I'll order up a personnel file on the mother.

Siddler had lived his life hard in the back streets of a dozen Asian cities. Sleaziness, he thought, thy name is Siddler! Jesus, he was sick of himself, sick of all the dope, sick of looking at dead bodies, ruined lives, Oriental cunts, smiling dink politicos with grenade launchers cuddled like babes in their smooth, brown forearms—sick, in short, of the vast and corrupt phantasmagoria of Asia that had nurtured him on its rich, brown paps and then cruelly cut him loose with little choice but to Serve His Country.

The promise of Siddler's childhood—the serene, sunny Malayan world of his Baptist missionary parents—had vanished in the Second World War. The young Siddler had since come far from his first wartime recruitment into the world of spies: he had come from idealism to boredom by way of resistance, revulsion, resignation, acclimatization, and delight—in that order. Killing bored him now. The killing of Luckett bored him. Luckett had been in some guy's coffin—some guy named DiMalco. Now that poor kid, where was he? Jesus. A few chunks and gobs of that guy, mixed with leaves and dirt, had probably vanished in the wrong green body bag somewhere out there in the combat zone and Luckett had somehow taken his place in the coffin. But how? Why?

Siddler pulled the crumpled cablegram out of his crumpled khaki fatigue shirt and looked at it again as the waiter placed another bowl of peanuts in front of him.

Not mothering much to go on. The basic facts of two dead. A couple of serial numbers, a few rough dates, the business about disfigurement and professionalism which didn't seem to make any sense, a suggestion to go to the morgue which for chrissake who wouldn't think of anyway, and a couple of not-so-very-well-veiled threats to get on the case in a hurry.

Siddler lingered over the words, "Typhoon brand." That was a new one. He wondered where that particular brand would come in from. There was nothing to do but work the case, work it as he always worked them—despite what those guys back in Washington might think.

He raised his chilled glass of blood-red liquid toward the rich greens of the golf course and paused an instant before drinking. "Good morning, Saigon," he toasted.

Half an hour later, Siddler walked into his tiny downtown office, squeezed past a pale, lean man bent over one of the room's three gray working desks, and sat down at his own desk.

"Ah yes, Ralph, I see you have survived the night," said the pale man in a high-pitched voice.

"You bet your ass I have survived." Siddler didn't look at the pale man, whose name was Trager. Trager was like the furniture. He came with the office, came with the job, got in your way mostly but could be of some utilitarian service from time to time. Trager was a narc, too, but mostly he was a supercilious sonofabitch. He always seemed to start his sentences with, "Ah . . ."

Siddler groped under his armpit and extracted a heavy .45 military semiautomatic pistol from a smelly leather half-holster strapped under the khaki shirt. He hated the weight of the thing there in the armpit where he sweated heavily, but there was nowhere else to keep a weapon of that size effectively concealed under summer clothes. Most people in Vietnam just carried their weapons on their hips in holsters, or they carried rifles. But Siddler had a psychological theory about that. The way he figured it, if you had it on your hip, they'd just lie in wait and ambush you. But if you had it concealed, they'd come into the open to get you and take their time. In those few moments, you could draw and gun them down. Siddler's 1912-model Colt took seven-round clips. In three seconds you could kill seven men with it. More, if more men happened to be standing behind the seven. Regular police .38s were for pussies, in Siddler's view. The bullets walked right on through whatever they hit, and whatever they hit kept walking right at you. That had never happened to Siddler, but he had heard stories about it. Siddler believed in firepower. He used soft-lead bullets and cut notches in them: dumdums.

He put the gun on his desk and rummaged in the trash there, looking for the additional package of information on Luckett that Gilmore had promised in the cable. He found it, tore open the heavy manila envelope, and pulled out several white, typed sheets of paper plus several photographs. Siddler examined the photos. There were three of them, all in color—an agency mug shot of Luckett, a full-length shot of Luckett's dead body, and a shot of Luckett with his new wife. Siddler examined the photo of the dead body. He had seen this sort of thing before, had seen the bodies of men after they had been tortured to death and disfigured in the jungle. He never got used to it, never completely. Christ, he wondered, what kind of pigs would have done it?

Next he looked at the agency mug shot, examining the smiling, clean lines of the young face. Just like my son, he thought. Not entirely true, he knew. His son had been half-Thai by a woman he had left more than a decade ago, and Siddler had lost track of both. He scanned the typed sheets. The kid Luckett was a newlywed. Gilmore had sent him over. Jesus, those Washington guys are trained in everything but common sense.

Not much in the reports. The kid seemed to have spent a lot of time at Tansonnhut. He had visited bars, the PX, the logistics areas, the South Vietnamese command centers. He had checked into the availability of black market rifles. Gilmore's summary contained basically the sparse information that Luckett's few reports to Washington had provided. Siddler read one item several times with interest:

JUNE 1: FINALLY MADE BREAKTHROUGH BUT NOTHING CAN BE DISCUSSED BY CABLE. IT WILL NECESSITATE TRIP BY ME TO THE SOUTH. I MAY BE OUT OF TOUCH FOR A FEW DAYS.

That was contained in the last report that Luckett had filed. Very mothering helpful, thought Siddler as he scanned it again. Jesus, don't those guys ever tell one another anything?

Siddler stuck the agency mug shot into his pocket and said to Trager, "I'd tell you how I survived the night, Trager but you never get out of the office."

The other man looked up from the pile of reports neatly stacked before him and smiled blandly.

"Ah, anything I can help you with, Ralph?" he squeaked.

"Yeah, what's this Typhoon brand heroin? I haven't heard of it. Maybe it's connected with this case I'm supposed to be working on."

"Oh yes," said Trager. Siddler watched him slide open a filing drawer in his desk and finger over the tabs on a row of manila folders. He extracted one and leafed through the contents.

"Ah, here it is," he said, scanning a page of single-spaced typing. "Typhoon brand heroin, number four, the purest there is—comes out of the mountains of Burma from a relatively new laboratory set up under the auspices of, ah . . . General Byprang, that's a Lao general slightly out of his territory. It is brought down from the mountains by mule train under guard of Nationalist Chinese guerrillas, also slightly out of their territory . . ."

Trager looked up for some sign of approval of his wisecracks but Siddler was waiting for more.

"Ah, well, the only thing different about this stuff is that it goes a new route, called the southern route by this writer here, about which not much is known since it's just been set up within the last year or so and only got fully operating during the last half-year, according to this skimpy report. The stuff does not come into Saigon by airplane from Vientiane as per usual. That's good, means we must have been having some effect by reining in the Vietnamese Air Force— but it comes in by land to the Seven Mountains area on the Cambodian border. At least that's what this operative says. Since the area is somewhat under Communist control, the operative thinks the North Vietnamese Army might have something to do with it. Nothing certain is known about how the stuff gets from there to Saigon, and all kinds of checks have failed to yield much more."

"What do the NVA get for it, rifles, huh?"

"I would suppose so, weapons of some kind." Trager looked up.

"Black market rifles," mumbled Siddler. It fit. Luckett had been looking into the black market in rifles. The trip south. Typhoon brand flooding Washington. It all fit nicely, but what the hell did it mean?

"Who's fighting for our side down there?" asked Siddler. "Any big-shot dinks down there?"

"Oh yes," said Trager.

He meticulously replaced the folder, closed the drawer, opened another drawer, fingered a similar row of folder tabs, and pulled one out.

He read in a monotone, "Lieutenant General Nguyen Duc Tran, commander of the glorious Republic of Vietnam forces in the Mekong Delta . . . ah, you'll have to excuse me, Ralph, but this is a news release from the information ministry . . ."

"Read on, I don't give a shit." Siddler was scribbling a few notes with a stubby pencil on a rumpled pad he kept in his back pocket.

"Ah . . . was born on July 20, 1930, in Vinh Hoi, Bac Lieu. He went to high school in Can Tho and Petrus Ky, Saigon. His military schooling includes the officer school at Dalat, armor school at Vung Tau, staff course at Hanoi, the advanced armor course in the U.S., the Command and General Staff College. His military career went as follows: 1950, cadet, National Military Academy; 1951, platoon leader, French Army—"

"Jesus, ain't there anything interesting in there?"

Trager scanned the sheet and turned the page.

"Well, ah . . . not much more, really. It says here that General Tran speaks, besides Vietnamese, fluent English,

French, and Japanese. He is married and has six children. He holds the black belt in Judo, plays an excellent game of tennis, and is an excellent swimmer."

"That's dandy. Anything else?"

"Yes, he makes his headquarters in Tinh Bien, down by the Seven Mountains. Ah, here are some photos . . ."

Siddler grabbed the sheaf of photographs and looked through them. A military portrait. Family shot. Candids: reviewing troops, looking through field glasses . . .

Jesus, thought Siddler, the little bastard is probably looking through those glasses at the girls undressing next door. Siddler had never met Tran, but instinctively he knew him, knew the swirl of corruption and death and intrigue in which the little general would live. If there were a new southern route for heroin through Tran's military area of operations, the odds were that Tran would be authorizing and perhaps even running the show. How else do these dink generals make it? Everything in Tran's delicately balanced world of political intrigue and military power would depend on getting large sums of money, and dope dealing was one of the easiest and most dependable ways to get it. With the money, a general like Tran could buy the treads to refurbish his battered tanks and the ammo to keep their guns blasting; he could pay the bribes that would keep food moving to his troops, buy the loyalties of a hundred village chiefs, and hold together the bloc of votes in the National Assembly that would make him a growing political power in South Vietnam.

"Lemmie see that file."

Siddler grabbed the folder from Trager and began leafing through it.

"Jesus yeah, I remember this guy," he said, pulling out a

faded clip from *Stars and Stripes* showing Tran sitting on the hood of his jeep, smiling and explaining to newsmen the success of his latest operation.

"TRAN BOASTS HIGHEST BODY COUNT," said the headline.

"That's the man who's planning to be president of this republic one day," said Trager.

"Yeah?" said Siddler. "Not if he's doing what I think he's doing and I catch him at it."

"Really, Ralph, you know you can't overturn the way they've been doing things for thousands of years."

"The Christ I can't," Siddler said. "You know if this guy associates with any white men in particular? I mean, any U.S. soldier types?"

"Not that I know of. Not that it says there."

"How about getting some checks in motion for us?"

"All right, but it's pretty sensitive and—"

Siddler cut him off. "My ass," he growled. "There are Americans getting killed in this rotting shithole, and that's the only thing that I'm sensitive about or you ought to be sensitive about."

He picked up the photo of Luckett's body and flung it at Trager's face.

Trager caught the photo and looked at it.

"My God," he whispered.

"You bet my God. You just order up some checks on that little bastard Tran. I want to know everything about him. I mean everything. I wanta know how many centimeters his mother was dilated after three hours in labor."

"All right, Ralph." Trager was subdued.

Siddler looked again at the photos of Tran. There was one in particular, a picture of Tran with friends in Saigon. Siddler studied the squat figure of the round-faced lieutenant general in his starched fatigues, spit-shined combat boots, and sun-glinting wrap-around dark glasses. The general was surrounded by half a dozen other people, and the group appeared to be chatting informally on the steps of some building.

"An excellent swimmer, huh?" he muttered. "I suppose he swims the dope back to Saigon."

Siddler handed the photos back to Trager, picked up his .45, and inserted it into the hidden holster under his arm. He left the top button of his shirt undone, as usual.

"Ah, what's it all about?" asked Trager, more careful now after Siddler's outburst.

"Can't tell you," said Siddler as he squeezed out the door.

It took Siddler fifteen minutes to get back out to Tansonnhut. The gate guards waved him through. Sweating in the hot morning sun and straining his beefy body upward to get high enough in the jeep seat made for taller men, he drove slowly along the base's dusty main drag.

Tansonnhut covered several sprawling square miles. It included a major civilian air terminal with commercial connections all over the world, an American military terminal with hundreds of daily flights all over South Vietnam, giant freight terminals, and endless acres of storage sheds, command complexes, munitions and oil dumps, rows of concrete hangars for fighter jets and bombers by the hundreds, troop quarters, amusements (including swimming pools), and post exchanges stocked with the best whiskey and other goods that affluent America could provide. Tansonnhut was a city complete in itself. In that swirl, Siddler knew, drugs and other contraband moved through in a thousand different ways to destinations around the world.

Siddler drove past the headquarters of the Military Assistance Command–Vietnam, called MACV for short. It was a three-story, prefab building that covered a city block and seemed to vibrate with the hum of the giant air conditioners that kept the high-ranking warriors cool inside.

He pulled up at the military air terminal and went in.

"What did he look like?" asked the counter clerk after Siddler posed his question about Luckett.

"Blond, lean . . ." Siddler handed over the photo.

"Christ, he could be any one of a thousand of these guys." The terminal was swarming with them. Men toted their duffel bags, attaché cases, suitcases, and rifles. Some slept on the floor or in chairs. Others sat dully, sweltering in the hot terminal, waiting to board the green-and-tan camouflaged C-121 and C-130 transport planes that would take them to their destinations.

"I didn't expect you to remember him personally," said Siddler wearily, "but how about looking him up in your paperwork?"

"Christ. All right. Hang on." The man disappeared into an office.

Siddler waited, leaning on the counter and looking idly across the room. He was starting to get a headache. He usually got one about this time of day. Maybe, he thought, he ought to cut down on the marys at breakfast and see if that would help.

Somebody was whistling a tune. When he heard it, Siddler's body stiffened.

> *And in,*
> *Her hair,*
> *She wore a yellow ribbon . . .*

A lean black soldier dressed in a class-A uniform and carrying a small suitcase was whistling as he walked across the room in long, easy strides. The man passed close to Siddler.

Siddler had seen that man before. He didn't know where or when, but from somewhere in his twenty-five years of rummaging through criminal dossiers, mug shots, and telephoto shots, he recognized that handsome, angular face. Siddler instinctively wanted to follow the man, but stopped himself. It was enough to know that the man was in Saigon. He could be located. There was no place in Saigon that Siddler couldn't discover and penetrate. Siddler made a mental note to run an all-agencies check on the man's characteristics in hopes of identifying him. Then he would go after him. Right now he had the Luckett case to worry about.

The odd encounter with the whistling man was the sort of thing that happened several times a week to Siddler. Locked up in his mind were thousands of faces, millions of facts. A chance encounter, a human quirk observed casually, anything could set the bells of recognition ringing. Then he would do what he always did: check it out.

The black man disappeared outside the terminal's swinging doors and the clerk returned with a batch of papers.

"Yeah, Luckett, one eight nine, two three, triple seven nine," he mumbled. "He came through, uh . . . arrived here May 15 and, uh . . . he apparently hasn't left yet, unless he went out on commercial."

"He went out all right, but it wasn't on commercial."

"Huh?"

"Never mind."

"He made, uh . . . a coupla trips in-country . . ."

"What trips?"

"He went to Tinh Bien on the, uh . . ." The clerk was rustling the papers as he studied them. "June second and again June the fourth . . ."

"Tinh Bien? You mean . . ."

"Yeah, you know, down by the Seven Mountains there in the delta."

Siddler thanked the clerk and left. Luckett had apparently caught on to something—the southern route?—in the Seven Mountains. Or had he? He might have seen the same report Trager had about Typhoon brand heroin, in which case any normal narc would go down to see what he could find. Luckett had found death, but had he found anything else?

Siddler drove to the morgue. It was a white, adobe-type building partially hidden behind green wooden military fences on a narrow street near the rear of the air base. Jets, heavily laden with bombs, thundered to takeoffs along nearby runways.

Siddler stepped over a small, scruffy dog who was peeing in the open doorway to the morgue's front office. He went through the office, where he found no one, and into a long, sunlit room with a row of cream-colored, corrugated tables down the center. There was a heavy smell of dust and formaldehyde in the air.

Several men in white smocks were working over corpses laid out on the tables. One of the men was a tall, graying colonel. Silver eagles were pinned to the shoulders of his smock. He was working intently over a corpse, brushing something on the lips with a tiny, white-bristled brush. Siddler came up on the other side of the corpse and stood, looking across at the colonel.

Finally Siddler said gruffly, "Hey, you the boss here?"

7

The colonel eyed Siddler evenly.

"I'm Colonel Lambert," he said. "Can I help you?"

Siddler explained what he was after and handed the photo of Luckett to Lambert across the corpse. The corpse was a young American man, open-eyed, well-muscled but overweight, with black hair. The man had been surgically slit in the normal Y-shape pattern for an autopsy. The abdominal cavity was empty. The man's internal organs were in a glass tub attached to the end of the table.

The colonel examined the photo of Luckett while Siddler examined the colonel's face. It appeared honest, unworried, and guiltless.

"No," said the colonel thoughtfully, "I've never seen this man. You say he was found in——."

"That's right, Colonel."

"Would you mind coming into the office?" Lambert led the way down the long room and into the small office where Siddler had originally entered. A large blackboard marked "HOLDING CASES" hung on one wall and contained a dozen names.

"I just can't believe it," the colonel said, turning the pages of a log book on his desk. "We're so *God* damned careful around here . . ."

"DiMalco. D-I-M . . ." Siddler was leaning over the colonel, trying to look at the log book.

"Here it is," said the colonel. "DiMalco, PFC Joseph V. Arrived on June fifth, done that night, went out on the sixth. Remains not suitable for viewing. Embalmer was Manes. It was all pretty standard, except . . . let's see, well, DiMalco was pretty badly blown up, from what it says here. Just bits and pieces. Weight, twenty pounds. That's all they found of him."

"Christ." Siddler was feeling woozy.

"Probably half of that was dirt and sticks. Manes didn't have much trouble with that one."

"Who's Manes?"

"Civilian embalmer under contract with DOD—Department of Defense. Works nights. I've got eighteen of them, plus six officers and one hundred and seven enlisted men. Then I've got six Vietnamese who clean up the building. I'll give you a personnel list, of course."

"Manes here now?"

"No, he's at home."

"Where's that?"

Siddler got out his pocket notepad and wrote down the addres Lambert gave him.

"Colonel," he said, "do you think there's any chance that our man Luckett got killed here—on these premises?"

Lambert thought about that one.

"Well," he said at length, "I guess anything is possible, but I'd like to think that it didn't happen here. I'd . . . well, I'd like to think it didn't happen."

"There's no chance of that, Colonel. You think the body came in here—but then why was it misidentified?"

"I don't know. That shouldn't have happened under any circumstances. I'll have a word with Manes."

"You and a few other people."

Lambert furrowed his brows. "Mr. Siddler," he said, "I

don't know what to tell you. There were a few others here that night, and you can get their names from this log book and talk to them, of course. But this morgue is tighter than a drum. I can tell you that because that's the way I run it. That's the way the President of the United States wants it run, and that's the way it is."

"Sure," said Siddler. Then he settled down to a series of questions that he knew he had to ask in order to satisfy Gilmore.

"Anyone come in here who doesn't work for you?"

"Only the MACV information people. Sometimes they come to get poop for their hometown news releases."

"Who comes?"

"Different people. And a man comes from First Log each morning to drop off the orders for the day's shipments. It's all routine."

"How does that work?"

"I can't cut orders here. I don't have the staff or the authorization. So we do it over at First Log—they send their man over."

Siddler stopped writing. "You'd better explain to me how this place works," he said.

Lambert led the way back into the embalming room and stood over several plain aluminum cases by a wall.

"The remains of deceased servicemen are returned from Vietnam in reusable aluminum transfer cases via military air to the Continental United States—to CONUS port mortuaries at Oakland Army Base in California and Dover Air Force Base, Delaware."

Siddler touched his toe to one of the cases. It moved slightly. Siddler had never toured the morgue before, though he had seen death from battle plenty of times, had seen the graves registration teams chopper in and set up their temporary five-corpse refrigerators at fire bases, had seen them stuffing bodies into the thick, green, rubber body bags. And Siddler had flown on choppers with stacks of bodies piled up like wood inside. He had watched bodies being unloaded, tossed to the ground like sacks of grain.

"There the remains are reprocessed, casketed, and shipped to their final destination." Lambert walked over to the corpse across which they had originally met. "The remains, including autopsied ones like Specialist Four Carter here, are embalmed in the normal manner using the arterial injection system and hypodermic injection of preservant formaldehyde solution. The bodies are then allowed to dry out, usually overnight. The bodies on the following morning are checked for preservation, the viscera or internal organs are returned to them, and the viscera are liberally covered with a paraformaldehyde hardening compound, following which the bodies are sewn and sealed with special mastik compound which prevents leakage—"

"How'd Carter get it?" Siddler wanted relief from the lecture, and turned to a human question.

"That's exactly what we wondered. Though he died in combat, there are no readily apparent wounds. Young Carter got it in the mouth with an AK slug that went between his teeth in an upward trajectory, never exiting from his head. Instead, the bullet went around and around inside his skull like a marble in a tub. That's how Carter got it."

"Christ," mumbled Siddler. He looked at the kid's face again. Then he looked at the guts in the glass pail. "That's what you use the hardening compound on?"

"Right," said the colonel. "It's a white, powder-fine sawdust compound which continues preservation action. We use anywhere from seven to twenty pounds of it. Then each body is wrapped in a plastic sheet which is overwrapped—you can see it over there—with a white or khaki cotton sheet and taped at the folds. The body is then placed in a polyethylene bag, air is evacuated from the bag by use of a regular vacuum cleaner, and the bag is sealed with tape. Then the body goes into the aluminum case which, as you can see, has no lining or other internal embellishments. The transfer case is then securely latched. The accompanying paperwork is placed in the dispatch tube there at the end of the case and locked. Oh, I forgot to mention . . ."

"Yes?"

"Because of the heat and the long flight out from Vietnam, my embalmers use about twice the amount of fluid they would use back home. Otherwise, our procedure is about the same except that the final cosmetology is left for the stateside morticians. We close all open wounds and do minor facial reconstruction work. You can see where Carter's lip has been sewn up there." He pointed. Siddler could see a thin, neat line on Carter's lower lip. "When they get them back in the States, either at the military morgue at Oakland or at Dover, or by a mortuary service contract, they completely reprocess the remains. You know, hypo the soft spots, reaspirate, and so on. Dress them up, so to speak, and then ship them to their final destinations."

"They couldn't do too good a job."

"What do you mean?"

"Well, they didn't discover Luckett."

"Why should they? There was nothing to look at. They probably just let DiMalco go through without checking, and I don't blame them."

A thick wooden door on one side of the room swung open and two men bearing a stretcher emerged. They moved to the center of the room, put the stretcher on a rack, and hefted a body bag off the stretcher and onto one of the cream-colored embalming tables.

Siddler could see beyond the thick, wooden door into the room-sized refrigerator where a dozen similar stretchers with body bags rested on white pegs protruding from the walls. Beyond that he could see outside through a second open door to where men in olive drab uniforms were unloading a mortuary truck, placing the bags on the stretchers. He saw one man lift a bag that was so light he picked it up with his right hand alone. The doors closed.

Near Siddler, one of the men who had been sitting along the wall waiting for work rose and approached the table where the bag had been placed. He had a big pair of cutters in his hand, and he went to work clipping open the bag.

"What's the first thing you do when you get a new stiff?" Siddler asked, notebook and pencil poised. Gilmore would want to know that.

The colonel winced but kept talking. "Identify him. That takes about half an hour. Usually there's no problem. When a lot of bodies are scattered around, the graves registration people give us a chart and tentative idents. Then we do it from fingerprints and dental records. When we've got a problem with idents, or when an autopsy is needed, we chalk them up on that board as a holding case. Otherwise, we move them through fast. A hundred a week, sometimes more. We don't like to keep them around. The President doesn't like to keep them around. Busiest morgue in the world."

A slim young black man began working on Carter's body. He stooped and turned a crank. The embalming table tilted gently toward Siddler. Blood and embalming fluid dripped into a metal trough that ran beneath the table. Siddler could see the fluids gather in a little stream in the trough and flow down a plastic tube that went like a gutter pipe from one end of the trough down to a large glass bottle fastened under it. The bottle was half-full.

"So what about security?" asked Siddler. "I could ship a thousand pounds of horse manure in one of those transfer cases and who the hell would know about it?"

"The MPs watch over the whole operation," said the colonel.

"They aren't worth shit."

"Maybe, but they check in here daily, seal up the transfer cases, and authorize movement to the aerial port. We load up the cases on our trucks and take them over to the cargo areas, a receiver signs for them, and they're on their way."

"How about escorts?" asked Siddler. "I once saw a body being escorted by a soldier, it was back in—"

"Remains aren't accompanied by escorts," Lambert broke in with his quick, crisp voice, "except in special cases where an escort is assigned to escort a specific remains. Son of a bigwig or something like that."

"How's it work otherwise?" asked Siddler.

"The morgue at Dover, or at Travis on the West Coast, would notify the special escort company when they had the remains prettied up," said the colonel. "The escort would meet the box at the airport and go with it wherever it was going. It would be in a new coffin, they would take care of that at the Dover morgue."

"It would go to the cemetery."

"Nope. Another morgue, near the parents or survivors. Then the escort and the survival assistance officer—the man who tells the family their man is dead—stick around until after the funeral. DOD usually gives the survivors a choice of two or three morgues in their area."

"That's nice of DOD," said Siddler. His eye lit on a cardboard box in the corner. It was filled with small plastic packets of fine white powder.

"What's that?" he asked.

"That's the hardening compound," said the colonel. "The embalmer will use it, for example, on these viscera here of Carter's." The colonel touched his toe to the box, which sat on the concrete floor. He picked up one of the packets and tossed it to Siddler.

Siddler squeezed it and sniffed it.

"Christ," he said, "it looks like packets of fresh scag."

Siddler headed back downtown. He badly needed a drink after what he'd seen. He sat down at the patio-bar of the Continental Hotel and ordered a gin and tonic. While he waited, he telephoned Trager and asked for some help. Then he enjoyed his drink and had another. Feeling better, he went to his jeep and drove to the address Lambert had given him for the Manes apartment. It was a three-story building with a large compound in the rear, located out in the direction of Cholon.

Siddler ducked in the front entrance and trotted up the stairs. He went to apartment B2 on the second floor, and put his ear to the cracked wooden door. He heard bed creakings and a low, sustained moan.

He tried the door. It was open.

Two figures lay on a bed, clinging together in the damp heat of the semidark room. The woman was moaning softly. At Siddler's step, the man turned and reached toward a table where a .38 caliber pistol lay with a watch, wallet, and loose pocket change.

"Forget it, pal." Siddler stiffened his arm and trained his .45 directly between the man's eyes. The man froze, look-ing at Siddler with hatred and breathing noisily through his nostrils. The woman lay back, naked in the sultry nest of the white sheets, her heavy body dark against them. Siddler could see that she was Vietnamese, though he guessed there was a good deal of French mixed in, judging by the European angularity of her facial features. Manes had a dark, handsome Latin muscularity. Siddler could smell their sweat in the humid room and he could smell something else—the heavy aroma of sex.

"What the hell do you want?" said Manes in a threatening tone.

"On the floor!" shouted Siddler, motioning violently with the .45 and advancing. "Move!" he shouted when Manes did not.

Manes eased up slowly, the physical manifestation of his recent endeavors diminishing rapidly, and lay down on his stomach on the concrete floor. He watched Siddler's right hand carefully, watched the tiny circle of the muzzle that never stopped pointing directly toward the middle of his forehead.

Siddler stomped his right boot on the small of Manes' back. Manes grunted in pain.

"All right," shouted Siddler, "what's your name?"

"Manes. Get your goddam—"

Siddler stomped savagely on the naked flesh and Manes grunted and whimpered.

"Listen you little bastard," shouted Siddler, "you talk back to me and I'll blow your brains all over this concrete floor. You understand?"

"Yes, yes," said Manes. "Now look, please—"

"Shut up!" Then, more softly: "You work down at the morgue?"

"Yes . . . ouch!"

The woman moved slightly, her heavy breasts rolling as

she lay back, the nipples vast splotches on smooth, dark skin. Siddler turned the .45 on her and grinned. She cowered back down and watched him intently.

"Stay still now," said Siddler to the man on the floor. "Just stay still, baby."

Siddler took two short steps to the night stand. He picked up the .38 and put it in his pocket. Then he examined the contents of the wallet. Manes had a Maryland driver's license but little else of interest. The license showed an address at a place called Hillcrest Heights.

"All right, Manes, you can get up now," Siddler said in a conciliatory tone. "Where you from in Maryland—where the hell is Hillcrest Heights?"

Manes got up gingerly and reached for a pair of trousers. He put them on and then sat on the other side of the bed from Siddler, near the woman. She looked up at his face inquiringly. She showed no fear, only interest.

"Near Washington, in the burbs," muttered Manes. "Now who the hell are you?"

"Siddler, Customs." Siddler stuffed his .45 back under his armpit and took in the woman's nakedness with relish. They come in good, neat packages, some of these dinks, he thought. This here's a beefy one, beefier than I've seen in a long time. "You did up a kid named DiMalco on June fifth. You remember it?"

Siddler lit up a Viceroy and tossed the match on the floor, watching Manes carefully for his reaction.

Manes looked concerned, tried to think. Suspicion registered in Siddler's mind.

"I don't remember," said Manes. "What's the problem, Officer?" Manes had turned cool, but it was too late. Siddler had seen—or so he thought—what he wanted to see: fear.

"I didn't expect you would," said Siddler offhandedly. "It was just another basket case."

"We get them all the time," said Manes.

"You work nights all the time?"

"More recently."

"Haven't I seen you somewhere?" It was a wild jab.

"Probably." Manes was sweating.

"Down at Vinh's, huh?"

"Maybe, I play there from time to time."

"But not since you started working nights?"

"No, not since I started working . . ." Suddenly Manes flashed with anger. "Now look, Officer, I haven't done anything. What's wrong?"

"Nothing's wrong, pal, except that DiMalco never made it back home. There was somebody else in that coffin."

"Impossible."

"Not so impossible. You ever see this guy?"

Siddler tossed the photo of Luckett toward the bed. It landed next to the woman, who picked it up and handed it to Manes.

"What's your name, honey?" Siddler said to the woman while Manes looked at the photo.

Her eyes hardened on him and she said softly, "Vu Thi Mai." The words, despite her hardness, had a birdlike, singsong quality.

"That's real nice," said Siddler.

"Never seen him," said Manes.

"Well," said Siddler, "that was the guy in DiMalco's coffin."

"DiMalco was in DiMalco's coffin," said Manes firmly, as if his words settled the question for all time. "Something happened somewhere."

Siddler snickered. "It sure did, pal, it sure did."

Manes handed back the photo. Siddler got up to go.

"What about my gun?" asked Manes.

"I'm gona take it, boy," said Siddler. "We're gona run some tests on it, then I'm gona send it back to you COD."

"But I—" Manes started to protest.

"What you need a gun for in this town?" Siddler asked as he went out, dropping his still-burning cigarette on the floor.

It took Siddler ten minutes to get back to his office.

He burst into the room and asked Trager, "So what number did the bastard call?"

"Ah, well, we didn't quite get the number because—"

"What?" Siddler exploded, thrusting his red, sweating face into Trager's.

"Now just a minute, Ralph," said Trager, holding up his hands. "We do know that he called out to MACV somewhere, somewhere on Tansonnhut, and we did get the conversation."

"Jesus, Trager," said Siddler, backing off, "sometimes I think you'd like to see me have a heart attack."

The two men bent over Trager's German tape recorder. Trager started in the wrong place at first, rewound it, and then started again.

Silence.

"Ah, here we go," said Trager.

Siddler listened intently. There were clicks and scrapes and a mild amount of static on the tape—then a buzz-buzz sound of a phone ringing.

The buzzing stopped and a voice said something that could have been, "Yeah?" or just a grunt.

Then, through a sea of static, Siddler heard, "Jesus, this is Manes," in a tone of high-pitched intensity. "The U.S.

Customs just sent an agent over here. I didn't tell him anything . . ."

There was a pause and a click.

"Hello," said Manes, "are you there? Hello . . . hello
. . . HELLO GODDAM IT . . . ARE YOU THERE?"

Then he said, "Oh Jesus," and hung up.

The entire recording took about ten seconds.

"Not much, eh?" said Trager.

"That's right, Trager," said Siddler. "Not too mothering
much."

He could tell nothing from the tape. Manes had made a
call after Siddler had left, but so what? Everybody in Saigon
had some sort of game on the side, and would be jittery
over such a visit. Manes handled bodies every day. It was
his job. Luckett might have been just another body to him.
There was nothing to suggest that Manes was implicated.
Whoever Manes called was too smart to talk over a telephone.

"Don't suppose you could bend a little and tell me what
it's all about," Trager suggested. Siddler gave him a quick
fill-in.

Then he asked for Trager's help in a number of jobs:
finding the most precise location possible for the number
that Manes had called; running a background check on
Manes and others on duty the night that Luckett went out
in DiMalco's coffin; checking with the MACV information
office and the First Logistical Command to see who their
liaison men with the morgue were; running Manes' .38
through the lab; and checking to see who the whistling man
at the air terminal might be.

"Sure, Ralph," said Trager. "I can do all those things for
you. What do you pay these days? Buck an hour?"

Siddler looked at Trager in mock surprise.

"Trager," he said, "you keep it up, boy, and you're going to be a human being yet."

A few minutes later Siddler again sat at the bar of the Continental and stirred a gin and tonic. A telephone had been conveniently placed on the bar in front of him and, beside it, he had placed Luckett's photo.

Siddler made several calls that yielded little or nothing. He called old friends and contacts. He called the MACV information office because the morgue colonel, Lambert, had said someone from that office occasionally dropped by the morgue. And he called the Tansonnhut headquarters of the First Log Command, the mammoth Army supply organization.

"First Log Shed Sixteen, Specialist Harris speaking, sir," a voice said through the static.

Static was always a problem when you called from the Vietnamese telephone system into the American military system, or the other way around.

"Yeah, hi," said Siddler. "This is Ralph Siddler down at the American embassy. Got a quick question; think you can help me out?"

There was a pause at the other end. Then Harris said, "Don't quite hear you, sir."

Siddler looked around the patio to make sure no one was close enough to hear, and then repeated what he had said, this time in a near-shout.

"I can try, sir," said Harris.

Though he tried for a confidential, soft effect, Siddler's voice rasped over the line.

"You fellows, uh . . . what do you do there in Shed Sixteen? What's that mean?"

"It's a logistics shed," said the faint voice. "It's the main office for First Log at the base. It's just a big shed full of stuff that comes in to be shipped out around the country. But I think you should talk—"

"You fellows got anything to do with that morgue?" broke in Siddler.

"Sir I think you should—"

"No, no, no," shouted Siddler. "Let me finish. All I wanta know is who is it from your shop goes over there to the morgue."

"Sir," said Harris, "I really think you should talk to Major Sims about this. He's acting duty officer now, and—"

"Now wait a minute Harris. Wait a minute. Don't get off the line. I wanta talk to *you,* Harris. Now just hang on there."

"All right," said Harris.

"You get that idiot major of yours on the line and I'm never gona find out anything, right?"

"Right. What're you, newspaperman?"

"Yeah, Siddler. *Charlotte Observer.* I don't want these guys to know I'm doing a story. Where'd you get your Ph.D., Harris, Princeton?"

"No, Midwestern. Political science." The two shared a laugh. Siddler choked on a mouthful of gin and tonic.

"What I need, here's what I need," said Siddler. "Who in your shop goes over to the morgue? Just tell me that."

"We share it," said Harris. "Must have been a dozen guys go over there. Nobody likes it."

"Nobody, huh? Tell me, Harris, you're a Ph.D. You know what an inkling is?"

"An inkling?"

"Yeah, an inkling. As in, I had an inkling you were a good man and would talk to me."

"Yeah. So?"

"Well, you just tell me Harris, just tell me this. If you had to pick a man and say you had an inkling, just the merest inkling, that he *did* like to go over for that morgue duty, who would it be? Think."

Siddler gulped his cool drink and waited.

There was silence on the other end of the line, except for the static.

Then Harris said, "You're not really a newspaperman, are you?"

Siddler said, "You bet your ass I'm not, sonny, and I'm not workin' for the gooks, either, so how about you gimmie a name."

Harris's voice came low and distinct and hard. He said, "Haverman."

"Wait a minute," said Siddler. "Lemme get my little notebook here . . ."

But Harris had hung up the telephone.

Siddler went to dinner alone at the International Club, then drove to Madame Vinh's. He parked his jeep in front of the large compound shrouded in thick trees. Madame Vinh's was more than a bar or a whorehouse or an opium den or restaurant or gambling house. It was all of these and it was more. The thing about Madame Vinh's was that you could get anything there that you wanted. Anything. You could even get a little kid if you wanted that, Siddler had heard, though he had never seen it done.

Siddler went into the small, first-floor reception bar. The room was dark, fish-netted, oared, and otherwise ornamented. If it had been in Georgetown or on Third Avenue it would have had sawdust on the floor. Crowded, sweaty men in fatigues, their weapons lying against their chairs, drank, played cards, wolfed down greasy hors d'oeuvres rushed to them by Vietnamese waitresses, stayed, raised, heard, clinked coins, cussed, grunted, farted. Siddler ordered a Beam, warm and straight, then stood fingering the shot glass on the dark wood bar.

Been one shitty day, he thought. Cream-colored tables bearing corpses floated in his Beam.

"Ah shore would lock to find me uh game," said an agreeable voice, and Siddler turned to behold a grinning face. "Ah shore would." The man said he was an engineer with an American civilian contracting firm that built most of the roads, bridges, and other paraphernalia that make modern warfare possible in a poor Third World country.

He could have been any one of thousands of pleasantly innocuous American civilians in South Vietnam. They were men who chose not to keep pace with the surging, affluent competitiveness of America. Here, on the edge of the Asian continent, earning twice what they would have at home and caught up in the confusion and excitement of war, they lived among a strange and fun-loving people whose police deferred to Americans and whose women loved them well. Here one could be free, here one could live and dream and pursue questionable pursuits unfettered by those stingy gatekeepers of the soul: wife, children, minister, neighbor, jealous co-worker, drainers-off of the juice and vitality of men.

Siddler regarded the engineer. My old friend! he thought. By the gods I love you well!

The engineer had stopped at a Cholon massage parlor the night before and had lost his wallet under amusing circumstances. His wallet had been in his trousers, his trousers had been around his ankles, his ankles had been under the bed, the bed had been against a wall, and the wall had a hole in it near the floor just large enough for a child to get through. The engineer, preoccupied with the intricacies of the massage being administered by a kneeling damsel, suspected nothing.

"Clever bastards," he summed up with a shake of the head and a smack of the lips. He and Siddler guffawed together and threw back their heads to down their liquor.

More liquor followed, and more talk. Patiently, they eyed the room for poker partners. There was no rush. Men were coming and going all the time. The evening was young and full of promise.

As if a thought suddenly fluttered to the forefront of his sodden consciousness, Siddler pulled out the photo of Luckett and tossed it on the bar.

"My son," he said. The engineer, elbows planted firmly on the beery dark wood, reached for the photo with a weatherbeaten paw. He gazed at it with nodding head.

"Good-looking kid," he said. "He a journalist too?"

"No, he's got some other kind of job, with some contractor, I think. I been trying to find him since I got to town. You seen him around?" Siddler forced the words out. Jesus, how he worked for a living! They should work this hard in Washington!

The engineer considered the photo more carefully, then let out a long "buuuuuuuurrrrrrrrrrrrrrrp."

"Don't reckon I have," he said.

Siddler pocketed the photo.

"He used to hang out with a kid named Manes, that's what he wrote me. You know anybody named Manes? Works out at Tansonnhut?"

The engineer said yes, he thought he might have known a guy named Manes, might have played poker with him right here at Vinh's. Then his attention drifted.

Siddler chugged his shot of Beam, swigged a little three-point from an English half-pinter to chase it, and framed another question. Did the engineer know a man named Haverman, a colonel, worked out at First Log? Siddler asked the question with apparent nonchalance as the engineer watched a scantily clad Eurasian girl unloading a tray of steaming spareribs, a process that involved her bending over and revealing the exact contours of her well-formed buttocks.

The engineer turned to Siddler.

"Shee-it," he said with disgust.

"What's wrong?" asked Siddler.

"You ask Chi-Chi what's wrong," he said. "Any friend of Haverman's is no friend of mine."

The engineer got up, picked up his drink, and walked off.

"Hey . . ." said Siddler.

But the engineer was out of earshot.

It was a bad case of the whirlybeds. A very bad case. At some point Siddler had telephoned Trager, but he couldn't remember when or why. He had tried to find Chi-Chi at Vinh's, but she had not been there and he had been told to return the next day for her.

Siddler rolled over, his head reeling with drink, visions of bodies and cream-colored tables floating in his mind. Why couldn't he get rid of them?

He didn't know why not. He just couldn't. The cream-colored tables swam murkily. For a horrible minute, he saw someone working over a body that was still alive. He looked closer, as if bending toward a television set, and saw that it was his own body.

He saw the mortician working slowly and carefully, patting the viscera with hardening compound, placing the viscera back inside the fat, beefy hulk.

He slept.

8

Vinh's was hot and stuffy in the late morning heat. Siddler went in past the little first-floor bar. He could smell the smoky dampness as he labored up the stairs to the second floor and down a long hall. At the end of the hall was a large, comfortable room furnished with deep cloth chairs and divans and carpeted in red shag. An air conditioner rumbled in one of the windows and the room seemed chilly in comparison with the dusty streets and damp hallway where Siddler had just been.

Siddler walked into the room. He saw Chi-Chi sitting in one of the cloth chairs, her legs up on a table—lean, graceful, serene. She was sewing, and when Siddler had stood there looking at her for several seconds, she looked up calmly and smiled.

"Hello, Mr. Ralph," she said in a small voice, a voice that made Siddler feel that she was glad to see him—and he knew right at that moment that he was no longer interested in what he had come for, that he could not conduct an interview, that he had thought and worried too much about the Luckett case so that his brain was fogged and he no longer knew what he was doing or what questions he should ask.

Siddler wanted a woman. He wanted Chi-Chi. It had been a long week and a long month and he had worked hard, drunk hard, played poker, sweated much—he needed a fuck, and he wanted to fuck this sweet little woman who was sitting before him with an almost tender smile on her delicate brown face, her eyes sparkling at him, her feet bare and small with red-painted toe-nails, her shoulders brown and soft and bare and curved nicely into the rough cotton cloth of her orange halter, her black slacks tight around her lean thighs. Women seldom guess how men want to fuck them like this, suddenly, how they are caught in a moment of hot desire that feeds on sweat and tropical sunlight and sudden dark cool rooms. Siddler, sucking in his breath, his chest going out, his gut sucking in, his face tightening into hard lines, his loins flashing hot and hardening right there as he stood before her, wanted her immediately, tenderly.

She saw the look in his eyes, and she softened. Siddler was no longer the shambling hulk. He suddenly felt himself a man. He might be gross and pocked of face—he knew he was—but in the tender embrace of this Chi-Chi, whom he had known and loved before, he was suddenly human and could even be—to her he guessed, hoped—beautiful, strong of muscle, beautiful with his white skin. She had said so once, and he had never forgotten it—"Ooh, ooh," she had said admiringly, awed, her eyes scanning his nakedness, ". . . ooh, beautiful, look . . . *beauuu . . . ti . . . ful* . . ." And she had run her hand over his whiteness and hairiness and manliness—and he had truly felt that she meant it, whether or not that was true, that she had admired something in him, had felt, and perhaps been the only person who had ever felt, that he was beautiful and perfect and human. He had paid her and paid her, and gifted her and souvenired her and had . . . loved her.

He moved toward her and touched her shoulder. She got up and kissed his cheek and led the way to the door on the other side of the room.

He followed submissively, his limp hand in her firm little hand, and he could smell the sweetness of her perfume. In the little bedroom, in the hum of its air conditioner, in its cool darkness, she took off her clothes first and then his, caressing him as she did it, seeming to dance before him in a little ballet of perfect nude beauty, her tiny breasts with their erect nipples bobbing, her body lean and brown and perfectly cool: she ruffled his hair, stroked him, kissed him —and Siddler, closing his eyes, returned her love, stroked her, put her down on the big bed's cool white sheets and made love to her until she came first and then came again and again with tight digging fingernails in his shoulders and hard little moans and cries and then Siddler came with a shudder, stopped, then pumped and pumped gently, until all their world seemed soft and cool and drained again.

They lay back on the sheets together and touched one another and smiled at one another. He asked more about her family and her friends, already knowing most of the answers. The gruffness in his voice softened. He wondered again whether he should marry her and told himself again no as he had done dozens of times before, knowing that it would sour and get wrinkled and bad if he did that, knowing that she did this with other men, too—but always telling himself that it could never be this good with the others, never allowing himself to realize her complete professionalism.

She was good. That was all he needed to know. She could love him, it could be true. That was all he needed, or perhaps all he dared. More than that was beyond his reach: stability and lastingness were beyond his reach or his competence, just as the white Red Cross doughnut dollies and other white women in Saigon were beyond his reach, the ones that all seemed to fuck for officers only, and so Siddler and Chi-Chi had moments like this, frail hours of time distilled from the frenzy of life and war all around them. It was very good.

"And do you know," he asked finally when he had given her money and when they were lying there having used up their conversation, "do you know an American colonel named Haverman?"

He was watching her out of the corner of his eye and so he was ready for her quick leap, and he caught her arm with his heavy paw and pulled her down again. She whimpered and averted her eyes, but he held her head near his and made her look at him. He asked again.

"It's important," he said, "it really is," and he kissed her on the cheek and the nose. "You must tell me."

Her voice was small as her hands implored his hands to soften their grip. She said in a soft, sad whisper, "Angela . . . die weef heem . . ."

"Here?"

"Yes."

"What do you know about him?"

"Nothing."

"Who knows him?"

"Nobody."

"Somebody does," he pressed. "Somebody must. *Please.*"

After a hesitation, she said, "Mr. Pei, he know."

"Mr. Pei in Cholon? The banker?"

"Yes, the banker."

He let her go then and she moved off the bed to her clothes. Siddler forced his hulk off the bed too and picked up his smelly garments from the floor. He dressed, pulling on his trousers, buttoning his rumpled khaki shirt, putting on his ungainly self again, his lumbering, sweating self, his realness. He bent over his belly to get at his shoe laces, straining. Her voice came again at him, unexpectedly, still a whisper, but fiercer, touched with fear and disbelief and despair:

"Angela . . . she waf eight years old . . ."

Siddler, his head throbbing in the afternoon sun, drove the jeep out through the few dozen downtown blocks of Saigon that still seemed shaded and almost serene, relics of the city he had known during the First Indochinese War —a city of colonial elegance marked by graceful embassies, ministries, great hotels, the presidential palace, the cathedral.

Now the war had transformed Saigon into a jumble of urban and suburban shantytowns flooded with refugees from the impoverished countryside. The streets were crowded with beggars, war-wounded, and draft dodgers. Saigon had become a disheveled slattern of a city—a national capital, yes, but also a focal point of competing international powers, a port city on the vast muddy Mekong River plain that was the political center of white men's hopes in Indochina.

Siddler drove into the cluttered morass of Cholon, the predominantly Chinese section. His senses focused on fish, because fish is what you smell when you drive out to Cholon. You smell fish in a thousand varieties and stages of decay. You smell the essence of Vietnam, its people, and its economy. Siddler's mind, which had been a blank until then, fluttered into a kind of rough focus and he told himself he was glad he didn't smell like a fish. He had the Western smell; his sweat and his stool had the red-meat smell, which is different from the fish smell. Siddler had been in Asia for fifty-two years, he mused, but it hadn't made a gook out of him.

His mind tottered toward other focuses, too, but some-

how missed them all. Luckett, Manes, Haverman, and that
whistling bastard . . . that sonofabitch, his name is Whistler, they called him The Whistler, and he's a low-level
drug pusher, a cheap dirty smuggler who isn't worth a
pound of that rotting carp out there. The identification
flashed through Siddler's mind and then departed: he
guessed he had known it all along. It meant nothing. Did
anything mean anything? Was there a theory of investigation that could tie all this craziness together? Could anyone
make a list of it all, and then check it all out?

That would be Gilmore's game. Neat cubbyholes. Follow-
through. Official inquiry. Paper churning. All horseshit. All
book-learned horseshit, and a hundred naive kids like
Luckett could be killed before one solid fact turned up. It's
no wonder the government is falling apart, sending half a
million kids at a time over to this country, all programmed
through IBM computers, shot up with a hundred doses of
the most modern medicines and ideas and military training,
and outfitted with modern camping equipment that would
turn L. L. Bean green with envy. A billion bucks worth of
flashy black rifles and ammo—but not one fucking grain of
common sense. And what had they done? Blown a beautiful
country to smithereens and killed and wounded what
seemed like half the human race.

Well, Siddler had a theory of investigation, too, in case
Gilmore or anyone should ever ask him. Once, when drunk
several years ago, he had formulated it for someone, and
the guy had laughed loudly and Siddler had dropped it and
never tried to explain it to anyone again. But basically it
was that you go to the important things and let the rest
flutter away. Let Trager handle the rest. Let the shit pile up
on your desk. Men, if they are men, have few needs and
strong ones. They need a woman to fuck, and sometimes

they need a lot of women: the bigger they are, the more women they need, which is good if you are an investigator because a woman will always talk, or can always be made to talk. And these strong men *need* money; usually they do what they do for money, and they always need a place to keep their money, and so acting on a good tip from a woman, you go to the moneyhandlers, which in Saigon means you go out to Cholon where the big Chinese mafioso chieftains of the black market are. And finally these strong men need psychological relief, they need friends no matter how big they are, they need to tell what they have done to someone, or to show what they have done, or what they are as human beings. They need to be loved or, failing that, to be respected. Or feared.

So if you think much at all, according to Siddler's theory, you think about cunts and dollars and psychological release. You go to the places that deal with those items, and you ask about a kid named Luckett who disappeared, about an embalmer named Manes who plays cards at Vinh's, and you ask why an otherwise perfectly normal, gentle, lovely prince of an American engineer would get up and walk out of a room at the simple mention of a man's name—Haverman.

"Yes, yes, Mr. Siddler," said Pei. "Come right in, I believe." Pei, ancient, wrinkled, white-bearded, with a shrewd glint in his black eyes, pronounced his "yesses" like "yisses." His voice was thin but strong: tensile strength like thin piano wire.

"Thanks, Pei," said Siddler. He hustled into the plain, cool room and sat down on a dark teak chair.

"Tea, I believe?"

"That would be great." Siddler wiped his sweating forehead on his sleeve. Pei instructed his white-suited servant to bring tea. The man scurried off and returned a few moments later with drinks.

"And how is your family, Mr. Siddler?" Pei settled himself in another chair and put his hands together so that the fingertips touched ever so gently, forming a little arch.

"Well, they're okay, I guess," said Siddler. "I haven't, uh . . . seen them for a while." Obsequious Siddler; groveling, fawning Siddler. Christ, come to the point. Remember the man is a chink. He may control a major bank and millions of dollars and half the black market in Indochina and a secret, private army of thugs that could bring down a small government—but he is still a chink.

"That's wonderful . . . I believe," said Pei. Only his lips moved. The rest of his face and body, including the finger-arch, were utterly still. The black eyes were a steely, flat, unreadable blank, as if a wire-mesh fence had lowered between the outside world and what was going on inside Pei's brain.

"And what can I do for you today, Mr. Siddler?" Pei, like most powerful men, knew how to gauge the value of lesser men. He dealt in human beings. Siddler was not someone he could afford to overlook.

Siddler showed him the Luckett photo and asked him about Luckett and Manes. Pei promised he would put out feelers. Then Siddler asked about Haverman, and in the same breath about Lieutenant General Tran.

Pei was silent a long time. Siddler watched him, watched as long as a minute, feeling something like electricity in the air as the old man's brain toted up the balances, checked the equations, and turned out the lists of what could and what could not be said. Siddler knew he would probably be used, precisely and carefully, in some grander scheme of things that he might never fully grasp. He didn't care.

"Well, Mr. Siddler, we have known one another quite a many years I believe, yes?" Pei's tone was softer than before and friendly. He smiled. "And I am certain," he added, "that what I would say to you here would never come back to me as to source, am I right, yes?"

"No sweat," said Siddler.

"Yes, right. Well . . ." and the tone became more serious, didactic, ". . . our friend, Mr. Haverman, I have heard of him, yes. I cannot tell you much, but as for his banking and . . . how do you say? . . . financial affairs, I can tell you a slight thing, as this gentleman has come to our notice."

Siddler's spine stiffened; he listened hard.

Pei continued: "Mr. Haverman has, from time to time, made . . . how shall I say? . . . deposited monies through our facilities here and in our affiliate banks in Hong Kong, monies I believe in quantities that are surpassing those that would be available to one of the rank, I believe . . . colonel, yes?"

"How much?"

"In excess, I believe, one million dollar in past six months," said Pei. "A little high for colonel pay, yes?"

"Yes." Siddler waited for more.

"As for Tran," Pei went on, "I cannot tell you much, but I can refer you to our minister, Lam. He will be expecting you in one hour." Pei scribbled an address on a piece of paper and handed it to Siddler.

Siddler marveled at the banker's neat precision. And at his apparent influence—he hadn't bothered to consult Lam in the matter.

"You may wonder why we are being so helpful," said Pei, "and there I can tell you so you won't mistrust us that . . . how do you say? . . . there are certain arrangements of ours that I am not particularly pleased to have General Tran interfere with. So we understand one another, I believe, yes?"

"And Haverman?"

Pei smiled broadly and made a graceful gesture with both hands as he shrugged his shoulders. "He is taking too much money, Mr. Siddler, it is as simple as that. No American should be taking that much money, I believe. I don't even know where he is getting it. I don't care. And now I must really be about my daily tasks. Good day, Mr. Siddler."

Siddler shambled across the shady square and across the dusty swelter of a traffic-jammed boulevard. Cars tooted and brakes screeched. He entered the three-story, tan stucco ministry building.

Inside, Siddler trudged up the blue cement stairs to the minister's office, wiping the sweat from his forehead with a khaki handkerchief that he jammed into his back pocket when he was through.

He entered a tiny reception room and gave his name to a male Vietnamese secretary who immediately ushered him through heavy wooden doors into a spacious room with a large fan turning slowly from the high ceiling. The secretary withdrew silently, closing the door.

Siddler stood blinking. The room seemed dark with its blinds drawn against the blazing afternoon sun and, for an instant, he thought he was alone.

Then he saw the minister, a tall, spare figure standing at a window on the far side of the room with his back toward Siddler. One of the minister's lean, brown fingers was delicately resting on a slat of the venetian blind and pushing the slat down just enough so the minister could peer out across the boulevard, over the iron fences and rolls of barbed wire, and into the green grounds of the presidential palace.

Siddler waited tensely. The silence, broken only by the gentle whir of the fan and by muffled noises drifting up from the street, lasted several minutes.

Finally the minister spoke in a soft purr, without turning from the window.

"Please be seated, Mr. Siddler."

Siddler walked forward two steps and sat down in a leather chair in front of the minister's desk. The minister pressed his finger down on the venetian blind and shifted slightly to look in a different direction.

Suddenly the minister let the venetian blind snap back into place with a metallic rattle and turned slowly until his watery gaze met Siddler's eyes.

Siddler could not help himself: he shuddered when he saw that face with its grotesque scar slashed down the right cheek, a sallow, drawn face under an uneven mop of greasy black hair, the mouth twisted to one side in unnatural chinlessness from war wounds of long ago.

The minister spoke in his purr: "I believe I might be able to help you in the matter of General Tran," he said, advancing slowly and sitting down carefully behind his desk. "Though . . ." and he paused to light a Gitane *filtre* with trembling hands. Siddler noticed that the hands did not cease to tremble once the match had been waved out, but rested uneasily on the desk. ". . . I am not pre*cise*ly sure that my little bit of information will fit into whatever pattern of investigation you—"

"Try it out on me," Siddler broke in gruffly, businesslike. They needed him. They thought they could use him. Siddler sensed this, and he pushed slightly. "I'm not so sure what the hell I'm doing, either."

The minister sucked and puffed. The cigarette trembled at his lips, in his fingers.

"It is a matter of shipments, Mr. Siddler," he said. "It is also a matter of a tiny hamlet called Chau Sit which, as you may know, is located two klicks northwest of General Tran's command center at Tinh Bien."

"No, I didn't know that."

"Yes, well . . ." Suck. Puff. Tremble. Siddler could smell the strong French smoke. He sneezed, wiped. Listened again. ". . . I think I don't need to elaborate. Am I right?"

"I think a little elaboration wouldn't hurt," said Siddler.

"The truth is that I don't know much more," said the minister. "Only that this weekend there is . . . expected another, you know . . . shipment."

Siddler got out his little notepad and stubby pencil.

"How do you spell that town, Minister?" he asked.

"C-H-A-U," said the minister, "S-I-T."

Siddler labored with the pencil. "Like set down, eh?"

Why were the minister's eyes wandering? Why didn't he look at Siddler? Did his hands always tremble? It was clear that the interview was about to end, and Siddler stood up to go.

"Thanks," he said.

The minister's watery eyes fastened on Siddler, then flickered away. He blinked rapidly.

"Thank you, Mr. Siddler," he purred.

"All right, Trager, paint a picture for me." Siddler threw himself down in the squeaking wreck of his battered office chair.

"Yes, Ralph, hi, how are you?" said Trager pleasantly, smiling up from his deskload of neatly stacked paperwork.

"I'm mothering wonderful," said Siddler. "What'd you

find out about Haverman? Then I'll tell you what."

"What?"

"Then I'll tell you about Haverman's sex life. Guaranteed treat."

"Ah, honestly, Ralph, since your call last night I've had a devil of a time, but I did get his two-oh-one file . . ."

"Read the mother."

Trager began in his slow, precise way to read from Haverman's military personnel file: "Colonel Rupert Kaiser Haverman, six seven two, three two, eight three one nine, forty-one years of age, eighteen years in service, four years in grade, infantry branch . . ."

Haverman, thought Siddler. One more goddam wasted fool of an American colonel in a pushbutton war. Siddler could picture him. He would be a physical sort of man—they all were. He would blend into a breed of American males who, whatever their other defects or accomplishments, were out to prove themselves and have their masculinity confirmed in the eyes of the world. In the oven-hot stillness of the Saigon afternoons, you could watch them running—one mile, two, five miles, around and around the little jogging tracks at Tansonnhut or downtown, pushing themselves, proving themselves. You could watch them, chests thrust out, arms swinging, striding along the streets, serious expressions stamped on their carefully suntanned faces, wearing freshly starched fatigues at two P.M., stopping to discipline troops who failed to salute them, forcing artificial gruffness into their smooth, middle-class, roast-beef-fed voices. Or you could watch them with their attaché cases going through the airport terminals and down the air-conditioned hallways of the command center at MACV, looking like so many executives rushing to catch a train.

These men, who appeared to be rugged individualists,

were mostly organization men at heart, timid and conformist underneath it all, more prone than most men to yield to peer and group pressure, men without much imagination but with a certain managerial talent that, when they got out, would land most of them jobs as supermarket managers, mid-level officials in obscure school systems, or maybe, if they were lucky and above the norm, even a vice presidency of a small manufacturing company.

"Soldier!" they would bellow at the slinking, recalcitrant youths of a generation they despised almost as much as they secretly despised themselves. "What do you do when you meet an officer?" Sneer: "Salute." Colonel: "SALUTE WHAT GODDAM IT?" Lip-curl, fang exposed: "Salute, *sir.*" Most of these colonels were decent, patriotic, and even sometimes selfless, but their big flaw, from a military point of view, was that they were managers at heart rather than fighters. Haverman would seem like them, Siddler imagined. He would blend right in. Who would ever know that there was that iota of difference—that he had killed an eight-year-old child?

Trager droned on: ". . . basic training at Fort Jackson, ah, sharpshooter's badge, advanced infantry training Fort Polk, machine-gunner's badge . . . Jesus, Ralph, you want all this really?"

"Keep reading, pal," said Siddler, "just try to skip some of the horseshit." Siddler put his feet up on the desk.

Trager went on, reading an account of a super-normal military career. Haverman had been commissioned into the infantry branch after attending officers candidate school, had attended the infantry school at Fort Benning where he became a paratrooper, and had then gone with a combat unit to Korea where, as a second lieutenant in charge of a platoon, he had distinguished himself under fire, winning medals for bravery. After that he had served in a number of routine assignments in different areas of the world, mostly Europe and the U.S. He had steadily but slowly advanced in rank. He had served mostly in field jobs that kept him out of staff offices, and during the last half of his career he had served in logistics posts. He was now serving his second tour in Vietnam, and he had an important job: officer in charge of the Tansonnhut depot of the First Logistical Command.

"And did you find out what the hell the First Logistical Command is?" asked Siddler.

"Runs all those storage sheds out at Tansonnhut. You know, the big aluminum jobbies that take all the air freight when it comes in and ship it around the country. A lot of heavy, tough work—all guided by computers. Haverman has an office in a place called Shed Sixteen. Now what about his sex life?"

"He fucked and killed a little eight-year-old girl."

"My God."

"Yep. A lovely guy." Siddler rocked forward and rummaged in the papers on his desk. "What's all this?"

"Ah . . . that's all the information you requested yesterday, Ralph. Tran backgrounder. Analysis of Manes's .38 —nothing from that. Manes backgrounder. Stuff on the others who were at the morgue that night. Haverman's medical records, his two-oh-one, some photos of him. A cable from Gilmore in Washington. There was one interesting thing."

"Yeah?" Siddler leafed through a report.

"MACV ran a routine check on that Whistler fellow, found that he's made eight trips back and forth to the States in the last year. Soldiers aren't supposed to do that, and they're after him."

"That means he's a courier. Eight trips, that's a lot."

Siddler looked at the stack of papers in despair.

"Jesus," he said, "you gotta be a Rhodes scholar to qualify for this job."

Trager smiled. "I was, you know, a Fulbright."

"Oh go fuck yourself."

Siddler dug through the papers until he found the photos of Haverman. He held up the most striking one and looked at it a long time—a four-by-five military portrait in full color. The face, heavy, square of jaw, muscular, savage, looked out at him with intense blue eyes from under a crop of crew-cut, bristling, rusty-red hair. Choler. Heat. The whole face gave off a reddish effect, came at you hot and pressing. Heavy lips. A strong chin: a dimple.

"A real charmer," said Siddler. "Did you look at this, Trager?"

"Looks like any other colonel to me," said Trager.

"I guess so," said Siddler. He tossed the photo down and rummaged in the papers. The stuff on Tran was boring to him. Siddler didn't care about politics, not beyond his vague, imperfectly formed perception that there had been some sort of political grand scheme of Pei's and Lam's into which he fit and from which he profited. They wanted to destroy Tran and Haverman, but they didn't want to be too direct about it. They hoped he would do the job for them. Lam, at least, appeared to feel guilty about it. Had they tried before and failed?

He opened Gilmore's cable and read:

URGENT THAT YOU CHECK OUT THE FOLLOWING AND IM-
MEDIATELY REPORT. WE ARE WORKING UNDER THE MOST
SEVERE TIME PRESSURE FROM THE WHITE HOUSE. AFTER
EXECUTION OF OUR INFORMANT HERE WE FOUND FALSE

MILITARY IDENTIFICATION AND MILITARY LEAVE ORDERS IN
POSSESSION OF HIS KILLER PLUS MILITARY UNIFORM. ARMY
ORDERS AND IDENT IN NAME OF PAUL FLYNN JOHNSON SE-
RIAL TWO FOUR THREE THREE TWO FOUR TWO THREE FIVE
PRIVATE FIRST CLASS. ORDERS DIRECT JOHNSON OF FIRST
LOGISTICAL COMMAND TO REPORT FOR DUTY IN SAIGON NOT
LATER THAN JULY FIRST THIS YEAR AND ARE SIGNED BY
ADJUTANT AT FORT MYER VIRGINIA. ORDERS CALL FOR
JOHNSON TO REPORT TANSONNHUT POST IDENTIFICATION
NUMBER US ARMY SEVEN SEVEN NINE FIVE EIGHT. IN
ADDITION WE FOUND CRYPTIC NOTE QUOTE WHISTLE BACK
SIX DASH TWENTY UNQUOTE. NEED EARLIEST HARDEST
WORK ON THESE LEADS FROM YOUR END PLUS IMMEDIATE
RESULTS OF YOUR WORK ON MORGUE AND OTHER LEADS WE
GAVE IN ORIGINAL CABLE. PERSONAL NOTE. I DON'T KNOW
YOU BUT MY FRIEND ROBERT HOLT SAYS YOU ARE BEST MAN
IN ASIA. PLEASE FOR CHRIST'S SAKE OUR ASSES HERE IN
MAXIMUM BIND WITH WHITE HOUSE AND IF NOTHING ELSE
PLEASE SEND US SOME HORSESHIT TO PLACATE THEM. WARM
REGARDS. GILMORE TASK FORCE WASHINGTON.

"Hey, Trager," said Siddler softly.

"Ah . . . yes?"

"Did you read this cable from Washington?"

"No. Interesting?"

"Yeah," said Siddler, "damned interesting. For the first
time it sounds like we aren't working in different worlds
from Gilmore and that task force in Washington. You got
that little book that gives military unit place designation
numbers?"

Trager clicked open one of his file drawers. "Right here,"
he said, pulling out a white pamphlet.

"Look up for me U.S. Army seven seven nine five eight,"
said Siddler. "It's somewhere in First Log, I think."

Trager thumbed pages, then ran his finger down the listing of numbers. His finger stopped, and he paused. "Seven seven nine five eight—that's First Log all right," said Trager. "It's Shed Sixteen."

Siddler sat in his jeep watching the main entrance to Shed Sixteen. He sat there sipping ice-cold gin and tonic from his canteen and watching soldiers go in and out, some walking, some driving forklifts and jeeps. His jeep was positioned alongside a mountainous stack of wooden crates containing airplane engines and behind a rigging shed. The shed had a roof supported by heavy timbers sunk in the ground, and no walls, so Siddler could see through it, through the cables and piles of goods and forklifts parked in it, to the entrance of Shed Sixteen, whose sheets of aluminum siding reflected the late afternoon sunlight with a hot glow. Siddler pulled his sweat-stained khaki campaign hat low over his eyes, slouched comfortably behind the steering wheel, and sipped the G-and-T. He thought pleasantly of the long cable he had composed and sent off to Gilmore, telling him about the morgue visit, about the possible significance of the fake orders that led to Shed Sixteen, about Haverman and Tran, about the presence of The Whistler in Saigon and how "back 6–20" could only mean that The Whistler would be arriving in the States on that date, probably at Dover, accompanying a shipment of heroin in some U.S. military conveyance.

Siddler had set Trager, who by this time had a firm understanding of the Luckett case, the task of checking military travel authorities to find out what flight The Whistler would be on. Then Trager would cable that information to Gilmore. Mere technicalities, thought Siddler. Mere clerical work. But Gilmore would be pleased. Siddler was getting a warmer feeling for Gilmore. The personal touch at the end of the last cable had been damned decent. They did need him, those smart-assed bastards, and at least they were man enough to admit it.

Siddler, sipping, didn't see Haverman until the big, red-haired man was almost directly on the other side of the rigging shed.

He was walking with long, muscular strides and wide-swinging arms, his uniform pressed perfectly, his boots clicking on the tarmac. He radiated purpose and vitality, and he was walking right past Siddler without seeing him.

Siddler, startled, sat abruptly upright, knocking his canteen to the metal floor of the jeep. The G-and-T spilled, gurgling onto the rusty metal.

"Oh shit," said Siddler.

Haverman turned in full stride, stopped with hands on hips, feet apart at shoulder distance, and glowered across ten yards through the open rigging shed at Siddler.

For that instant, the two men's eyes met.

Siddler, staring into the eyes of the real article, the living, breathing reality of the color photo he had seen earlier, saw Haverman start slightly with what Siddler took to be shock and perhaps even—he dreaded the thought—recognition. It was possible. Manes could have tipped him off, if there was a connection there.

Haverman wheeled and continued walking in his original direction as if nothing had happened. The whole episode had taken five seconds.

Siddler watched him go until he disappeared around the corner of Shed Sixteen.

Then Siddler gazed forlornly at the canteen on the metal jeep floor, its rust stained by the lost cocktail. "Jesus," he mumbled. "Two big mistakes in one short minute."

9

It took Siddler only half an hour of telephoning the next morning to confirm that he had made another mistake. He had failed to keep track of Haverman, and it turned out that the colonel was accelerating his pace.

The Land of the Free helicopter pad at Tansonnhut confirmed that Colonel Rupert Haverman had taken a flight to Tinh Bien near the Seven Mountains at dawn. The clerk noted, in looking at the colonel's reservation sheet, that the flight had originally been scheduled for Friday—two days away—but that the colonel's aide had called last evening and changed the date. The clerk also noted, with some embarrassment, a scribbled notation at the bottom of Haverman's reservation sheet saying that the flight was TOP SECRET and the fact of Haverman's going to Tinh Bien was to be disclosed to no one. The clerk had not seen the notation until too late.

"That's okay, pal," said Siddler. "I'll have Uncle Ho get a personal thank-you note off to you right away, and I'll tell you what."

"What?"

"You tell anyone I called, especially Haverman, and I'll have your stripes and your balls—in reverse order."

"Yes sir."

Siddler hung up and looked at Trager. Then he looked at the pile of junk on his desk and at his hairy hands. He stared a long time at his rusty .45 resting on some papers in front of him, and finally he said, "Shit."

"Ah, got a problem, Ralph?" asked Trager.

"That's right, Trager," said Siddler. "I got a problem."

Trager's approach to life never ceased to mystify Siddler. How do you tell a guy like Trager, who never lifts his ass out of an office chair, what it is like physically to chase a man? How do you explain that it makes fear well up in you and makes your knees shake, literally makes your knees shake, to keep after a man who at any instant could stop and turn and hide and wait to blow you away? How do you tell that guy about the realness of a bullet, its solidity, speed, and searing heat?—not just something on paper that Figure A is doing to Figure X in Diagram Y, but a real honest-to-God bullet fired by a man who wants to do just that? How do you tell a guy like Trager that the Haverman game has all of a sudden become a different kind of game, that it has stopped being just "a nice surveil and checkout"? How could you convey to a man like Trager what it felt like to look into those eyes of Haverman's and to sense that they recognized you, to sense that at that instant Haverman had become infinitely more dangerous?

"The problem is," Siddler finally said, "that Haverman left town today at dawn . . ."

"Ooooooooh . . ." said Trager. The rational mind. Click, click, click . . .

". . . for the Seven Mountains and Tinh Bien . . ."

"Hmmmmmmmm . . ." Grave consideration.

". . . and that he moved the date of his trip up after he spotted me surveilling him last evening."

"Geee-*sus,*" Trager breathed.

"Watch it, Trager," said Siddler. "Mommy will spank if you use words like that."

"Ralph, that *is* bad."

"Yes, it *is* bad, Trager," said Siddler, mocking his companion's solemnity. "It is mother*fuck*ing bad, *'deed* it is."

Trager thought for a few moments and then said precisely:

"It means: (*a*) he's guilty of something big; (*b*) he's clearing out faster than he had intended to; (*c*) whatever he's going down there for, it's damned important, and he's willing to risk everything for it; and, (*d*) if we're at all in the ball park, it's a shipment of heroin."

Trager looked at Siddler with a slight analytical smile: thin-lipped, self-satisfied.

"You forgot just one thing," said Siddler. "You forgot item *e*."

"What's that?"

Siddler picked up his .45, checked it, jammed in a clip with a metallic snap, stuck it into the holster under his arm, and got up from his desk.

Then he said, "Item *e* is that he's dangerous as a motherfucker, that somebody's gotta go after him, that there are two of us, and that you ain't the one's gotta go, you chicken-shit narc."

Siddler looked down from the windy, swirling world of the Air America helicopter at the hazy expanse of the Mekong Delta, a dainty latticework of light tans and greens. Most of the endless rice fields below him were irrigated to glassy sheets of water by hundreds of crisscrossing canals connected to the Mekong River. The river itself, whose tributaries begin a thousand miles to the north in Tibet and China, is muddy and languid as it crosses the border from Cambodia into South Vietnam not far east of Tinh Bien and then meanders to the bright, sparkling expanse of the South China Sea.

Siddler looked out across the plain, where half the popu-

lation of South Vietnam lives. He knew that men had culti-
vated poppies in the mountain meadows of central Indochina for thousands of years, transported their bundles of
opium down jungle trails by pack train, and turned the
balmy nights of their existence balmier with bittersweet
smoke. But it had taken the modern world, with its chemistry labs, to process the opium into a fine white powder
called heroin, and with its efficient international transport
system, to pay a man half a million or a million or more
for playing a part in the drama. Men have always fought
over the precious opium, but it had taken the modern
world to churn it out in megalots that can destroy not only
individual souls but whole cities and nations.

At Tinh Bien the flatness of the great plain is broken by
a cluster of rocky mountains, ten miles long and five wide.
Siddler had seen them once before. These Seven Mountains, or Seven Devils as they have been called during the
decades that armed men have clashed and died on their
jagged breasts, seem to brood evilly over the plain as you
approach them from the air. They look huge. But if you
are standing in the dust and sweltering heat of the American
compound at Tinh Bien, you get a different idea of them.
They seem like big rock piles, dark and dirty green, spotted
with scrubby bushes. Here and there groves of palms stretch
down from the bases of the mountains and dissipate on the
plain. The Vietnamese town of Tinh Bien itself is a series of
thatched huts, topped incongruously by a forest of television
aerials, and French colonial villas strung along the Vinh Te
Canal, which marks the border with Cambodia.

Nobody lives on the mountains but the North Vietnamese
Army. Several thousand battle-hardened troops live there
like rats in a network of caves beneath the rocky surfaces.
South Vietnamese farmers live in towns and villages and

hamlets around the bases of all the mountains, sometimes only a few hundred yards away from these caves. The farmers fan out into their fields at daybreak and return in the evening. When night falls, the soldiers in the mountains come down to attack military installations like Tinh Bien's American compound and the series of outposts strung along the border to keep the NVA from bringing in supplies from Cambodia. You can see their flashlights moving on the dark mountainsides at night.

No one is ever sure what the NVA troops are going to do. Major attacks are always expected. One thing is sure: the NVA will stay in the mountains at any cost. The mountains are one of the few natural refuges in the entire delta, and they are the major staging area for the movement of troops and supplies to NVA and Viet Cong units throughout the region. To try and assault the troops and drive them out of the mountains means only to fail and to die. The allies gave up many years ago trying to do that, contenting themselves with patrolling the lowlands around the mountains and trying to prevent border crossings.

The next best thing to do is to make life unpleasant for the NVA soldiers living on the mountains. This is done. Artillery shells rain on the mountains continually, night and day. B-52 bombers from Guam and Thailand come over nightly and drop their gorgeous turds. Jets sweep in to pinpoint-bomb the mouths of the caves, often dropping napalm. Helicopter gunships buzz like flies over the mountains, spewing rockets and God knows what else. The Americans dropped thousands of time bombs on the mountains. Some go off the next day, some the next week, some when you step on them and some when you fart within five yards of them. Is it any wonder that trees don't grow very well around here? It is a very unhealthy environment.

Men who live under these conditions, when captured, will tell you that a man's ears spurt blood when a 750-pound bomb from a B-52 goes off ten yards away outside the mouth of his cave. They will tell you that they have learned to keep their mouths open during the bombing, because if you keep your mouth shut at such a time the pressure from the exploding bombs will explode your brain, too.

Siddler walked as quickly as he could, his belly bouncing. He walked away from the small helipad where the flapping rotor blades of the chopper stirred a dusty whirlwind. He grasped his knapsack in one hand and held his other hand to his campaign hat to keep it from blowing off. The two canteens hooked to his belt gurgled and slapped with his lumbering motion. One was filled with gin, the other with tonic water.

Siddler went up the dusty road, past barbed wire, sandbags, machine gun and mortar positions, American and Vietnamese soldiers. The Americans were lying in the sun getting tanned. The Vietnamese, cleaning rifles, looked up and grinned at Siddler. A mimic, he grinned back. Frigging dinks would just as soon shoot you down as grin at you, he thought. He asked one American, a cook who was emptying coffee grounds, where he could find the commander. Siddler followed his directions across the compound. He went down a short set of wooden stairs into the sandbagged Tactical Operations Center, the TOC, which was half underground.

"WHEN YOU'VE GOT 'EM BY THE BALLS, THEIR HEARTS AND MINDS WILL FOLLOW," greeted a black-lettered sign on one wall of the TOC.

"Who's in charge here?" growled Siddler at a man bending over a map table in the center of the bunker's main room.

"I am," growled the man, wheeling around to face Siddler. "Who the fuck are you?"

Siddler dumped his knapsack in a corner. He unhooked the two heavy canteens from his belt and tossed them on top. Then he flopped down in a chair.

"It's hotter than a motherfuck in here," he said. "You got any ice?"

"*I said who the fuck are you?*" shouted the commander, slapping the map table.

"*I said it's hotter than a motherfuck!*" shouted Siddler.

The commander, who wore silver eagles on his shoulders and the name WATSON on his tag, looked in apparent disbelief at the fat little man.

Siddler looked at Watson, too, looked at him carefully because he knew that he must quickly take the measure of this man who must be trusted to help him and not betray him. Let Watson make a wrong move, better for Siddler to find it out now before he confided his mission. Better to put the man under pressure now on an irrelevant matter. He was here to ask about a colonel named Haverman, a hamlet called Chau Sit, and a shipment that could be heroin. For all he knew, Watson was involved in whatever was going on. Siddler had only a few minutes to test Watson's reactions. If Watson didn't pass, Siddler would walk out. He had done it before.

Watson laughed. *Three points* on Siddler's ten-point scale of good behavior. Desperate men can rarely see the humor in a situation. Watson's short, brownish hair waved slightly in the blast of air conditioning from a window. His smooth, boyish face broke into spreading crinkles of mirth, crinkles that went naturally in the smile way and not in the frown way. *Four points:* men who play for the big stakes frown more than they smile. Siddler remained sitting and looking up insolently.

"You got ice?" repeated Siddler.

"Christ," said Watson, "for a spook we got anything." *Five points.* Not a wrong move yet. Watson motioned to a sulky PFC to fetch ice and cups. "I shoulda known you'd be a spook, coming in one of them Air America CIA birds. Welcome to Tinh Bien, my man."

The ice came in pint-sized metal canteen cups. Siddler mixed two giant G-and-Ts and handed one to Watson. The PFC looked on with a scowl. Watson told him to get lost.

Siddler drank about half his cup in one breath—drank it like water. He let out a long, satisfied, *"Ahhhhhhhhh,"* smiled for the first time at Watson and said, "Siddler, U.S. Customs."

"And by God it is hot," said Watson with locker room friendliness. "That damn air conditioner just don't put out like it used to and it's harder than hell getting supplies down here in gookland." *Seven points.*

"What's all that shit?" asked Siddler, waving his hand around at the array of equipment in the room.

Large, plastic-coated war maps with arrows and other symbols drawn on them in various colors of grease pencil covered most of the wall space and were strewn over three tables in the center of the room. Low doorways led to dim rooms off the main one and the rooms were all filled with the low cackling of radio sets. The far wall from where Siddler and Watson sat was covered with stacks of electronic boxes with hundreds of dials and screens. A black American soldier, dressed in a white undershirt, camouflage fatigue pants, and combat boots, sat in front of the boxes and watched blips and lines cavorting on the screens. Watson explained that a blip on one of the screens might indicate that a hidden sensor along the Cambodian border was being disturbed, possibly by a troop movement.

"Or by a cow taking a piss," he added, "which is usually the case, and in which case the cow is in some very deep shit because we will probably put an artillery strike on his ass just in case."

Ten points.

"Watson," said Siddler, "if you're wondering why I'm here, I'd like to tell you in . . ." He glanced at the soldier watching the radio sets. Watson got rid of the man while Siddler poured a new drink for himself and filled Watson's cup to the brim, and then they talked alone.

"I'm after a colonel named Haverman and I haven't got much time."

"Yes, he came in this morning."

"Where is he?"

"Don't exactly know. He comes in here often, does logistics work. He got a jeep this morning, then took off. Probably went down to General Tran's."

"I'm after Tran, too."

"Good. He's a shit." Watson explained how American combat forces had left the delta as part of the Vietnamization program. Watson, like most of the other Americans remaining in the delta, was in an advisory role. Other Americans, like Haverman, handled logistics, most of the airpower, and other matters. But the ground combat was entirely on the shoulders of the Vietnamese who were under Tran's command.

"Does he smuggle dope?"

"Probably," said Watson. A serious expression wrinkled his face in a different way. "I honestly don't know. I'm

supposed to advise him, get him air support, and so on—
but he steers clear of me when he can. My problem with
him is that I want to do a hell of a lot more bombing over
there in Cambodia than he does, and I frankly suspect that
it's because his brother may be a high NVA commander
in these Seven Mountains."

"You can bomb in Cambodia?" said Siddler.

"Sure," said Watson, "we do it all the time, when that
little bastard Tran will let us. One of the Seven Mountains
is over there on the Cambodian side, and it's their command center. Honeycombed with caves, bristling with electronic communications gear. That mountain is crawling with
gooks and always has been. You can stand on the goddam
front steps of the mess hall here in the compound tonight
and watch the goddam truck convoys of NVA supplies—
you can see their headlights, can you believe it? Coming
right at you down from the north until it seems like they're
gona drive right in the front door of the mess hall and have
dinner with us. But then they make a sharp turn and go in
toward the mountain and disappear. So I ask Tran if we
can't bomb them regular-like, and he's always hedging,
always cutting it down. Christ mothering Jesus. You'd think
the truck convoys were bringing in supplies for *him*."

"Maybe they are."

"Wouldn't surprise me," said Watson. "I swear to God
it wouldn't."

"What's the relation between Haverman and Tran?"

"Not sure. They smuggle dope together for all I know."
Watson sipped his drink and then said earnestly, "I'm sorry
I can't tell you more about Haverman. He comes down here
every month or so. He gets me some supplies, but it's all
routine. Mostly he coordinates with Tran's headquarters.
That's where he goes mostly."

"So if they were smuggling dope together, it would figure."

"Well, yes," said Watson, "but I've never had any indication that they are. It's his job to supply Tran. Maybe that's all there is to it."

"Ever hear of Typhoon brand heroin, or something called the southern route for getting it to Saigon?"

"Vaguely. You gotta remember I spend most of my time fighting a war down here."

"Ever hear of a little hamlet called Cow Shit . . ." Siddler fumbled in his back pocket for his notepad.

"Chau Set," he said. "C-H-A-U and then S-E-T."

"Sit," said Watson, "it's Chau *Sit*. S-I-T."

"Right, right," said Siddler, examining his notebook more closely. Then he demanded, "What's out there? What's at that place? Where is it?"

"About two klicks from here. It's just a hamlet. We've had a lot of action in there. I guess it's a VC hamlet, though it's rated one step above that on the hamlet pacification survey—for appearances."

"The Viet Cong control it?"

"Yep, except in dazzling daylight. You wouldn't want to go in there for a moonlight stroll, and we've taken fire in that sombitch sometimes even during the day. You're standing there minding your own business and suddenly the bullets are kicking up the dust all around."

Siddler was puzzling it out in his mind.

"If they're shipping heroin," he said finally, "I think they'd be making the transfer there." He looked steadily at Watson, who shook his shoulders slightly.

"Ain't nobody in that hamlet but the commies," he said.

"You said Tran's brother is a Communist commander."

"That's right."

"Then they could be getting it from the commies," said Siddler.

"Yeah," said Watson, "and paying for it in guns that Haverman can get easily and that the dinks then use to shoot at me. Son of a *bitch*. Wouldn't surprise me in the least. But why Chau Sit? What makes you so sure the shipments go through there?"

"A tip," said Siddler. "Can't tell you. High, good source."

Siddler tossed Luckett's photo to Watson. "Ever see this guy?"

"Yeah."

Siddler's head popped up.

"Here? When?"

"Oh, few weeks ago, I don't exactly remember. Came through saying he was a newspaper reporter. I talked to him for a while, gave him some bullshit about how well we're doing down here, and sent him off to chopper around a little. Never heard from him again."

"Where'd he chopper to?"

"Oh, Tri Ton, Olam . . . and Jesus, Chau Sit."

"He was an agent of ours," said Siddler. "After Chau Sit, his next stop was a coffin." He slugged down the last of his drink and stood up so he could get a better look at the wall map. "Show me where that hamlet is, will you?"

Watson took a pointer from the table and indicated a spot on the map.

"Right there, right close to the border," he said.

"Where are we?" asked Siddler.

Watson moved the pointer a little south. "Right here."

"And Tran?"

Very tiny pointer move. "Here."

"All right," said Siddler, "you got troops?"

"Now just a minute," said Watson, his voice rising. "I got a war to fight around here and I don't have time to go flying around the wheat fields—or whatever the hell they are—chasing after two-bit dope smugglers."

Intimacy, and then the power grab, the jostling for position: it always happened that way, in marriage, in friendship, in everything. Siddler was ready for it.

He put his cup on the table, turned full-face on Watson, and spoke gruffly: "Colonel, I'm gona tell you a thing or two. We're not after two-bit dope smugglers. We're after some of the largest dope operatives in the world, and what they do affects the national security of the United States. It *is* . . ." and Siddler paused to catch his breath and dig deep in his throat for his growlingest bulldog tones, ". . . the very significance and magnitude of my mission here today that prevents me from fully disclosing it to you right now . . ."

Watson's cheer had left him, but so had his anger. He was obviously taken aback by Siddler's unexpected offensive and its ring of authority.

"And Colonel, if you doubt what I say, if you doubt that I have need and authority to request your services and aid, then you pick up that telephone and call the White House in Washington and talk to your commander-in-chief and mine, the President of these United States, because mister . . ."—and Siddler again paused to catch a breath and wipe the sweat from his face, which was red from the magnitude of his sudden rhetorical effort—". . . *that's* whose direct motherfucking authority I am operating under here today, *that's* who sent me a cable containing my order to come here through the American embassy in Saigon this afternoon, and *that's* why if you're thinking of lollygagging or not pitching in and giving every goddamned ounce of help that you can give me, then mister you better pack up your gear and twaddle right out of here . . ."

It petered out. It had to. Siddler couldn't think of anything else to add, and he had already missed half a dozen good chances to stop with the proper flourish.

Watson was silent for a few moments.

Then he said, "All right, Siddler. Name it."

"Get a man on that sombitch Haverman right now," said Siddler. "I don't want him to fly the coop. Just have someone surveil him nice and easy, and let us know if he looks like he's gona leave town."

"All right."

"Our next problem is a time problem," said Siddler. "This shipment, my informant says there's an imminent shipment. If they're getting it through Chau Sit, then when would it be? Tonight?"

The prospect of doing anything tonight but getting drunk was highly unsavory to Siddler. He already felt that he was doing more mental work than he had since high school. His head felt lighter from the drink. He wanted the cares to drift away, wanted that hateful precision of mind to leave him.

"Could be tonight," said Watson, "but there's an operation tomorrow and—"

"A what?"

"An operation. You know, thousands of troops sweeping, air strikes, artillery, we chase the little bastards around and try to corner them and spend millions of the taxpayers' dollars and it's the biggest fucking waste in the world. Starts tonight with an Arclight. You'll love it, Siddler. You'll see some action."

"An Arclight?"

"Sure, B-52 strike. Best show on earth. You'll see it tonight. It'll blow your mind. Then we move out at dawn."

"Wonderful," said Siddler.

"Holy shit," said Watson, struck by a sudden thought.

"What?"

"You asked about timing, and I just remembered that the operation wasn't set for tomorrow until last evening. It was set for the weekend. Tran called me personally last night and moved it up. Said he just received fresh intelligence."

"What time did he call?"

"Six-thirty."

Siddler's mind, though becoming glazed, could still calculate that there was one hour between when Haverman spotted him at five-thirty and when Tran had called Watson at six-thirty. It fit. The only trouble was, Siddler still wasn't sure that the pieces were fitting in the right puzzle, that he wasn't completely on the wrong track. He was beginning not to care.

"All right," he mumbled, "that's when they're going to do it and we'll get us some troops and some choppers and get out there at dawn and surprise them when they come for the heroin."

Watson shook his head. "It's very tricky," he siad. "More likely they'll surprise us. Ain't no way we're going in there without them knowing about it."

Watson, using his pointer, outlined the next day's operation. A big one. Battalions sweeping the plains, sweeping around the mountains. Driving the enemy, supposedly, in certain directions. Driving him into blocking forces waiting in ambush positions. Driving him into the wide palm grove aprons of the mountains, then holding back while artillery pounded him, while Puff the Magic Dragon—a huge C-130 transport equipped with miniguns firing tens of thousands of rounds per minute—raked him over.

Too bad it didn't always work out, said Watson. Usually the operations ended up destroying trees and monkeys but few NVA at enormous cost. Tomorrow would be typical, the kind of operation they had every week or so. Usually the enemy knows the details beforehand and manages to stay the hell out of the way. "Wonder why? I think I know why. I tell that little mother Tran everything, he tells his men, and the word goes out. Just like that."

As Siddler's head began nodding, the room was suddenly filled with a splattering roar: the evening monsoon on the tin roofs of the compound. It had been sunny and clear fifteen minutes earlier. Outside, men rushed for cover as the sheets of rain drove down from black clouds racing across the sky. Dry, caked earth instantly ran with little streams.

Watson and Siddler looked out a tiny window across the wet plain. Secure and dry in the command center, they watched the thunderheads rolling across the sky. In the distance, they could see peasants still working, hunched figurines, far away in their wet fields.

"A thousand years," said Watson. "A thousand fucking years these gooks been tilling these plains, killing each other, and we think we're going to have some effect on them with this Arclight tonight and this operation tomorrow."

He shook his head slowly. They watched the rain blasting down.

"Ain't no way," Watson said. "Ain't no fucking way."

Siddler sat alone at the bar in the tiny officers club and listened to the throb of fifties nostalgia stuff on the juke box. He watched a few troops playing poker, fondling Vietnamese girls, and drinking as much as they could get down.

Watson had gone after dinner to meet with a Vietnamese province chief, or something, leaving Siddler on his own. Siddler was getting drunk and having his doubts about the next day's operation.

Why didn't he trail after Haverman directly? Why take a chance, a wild chance really, that Haverman would show up at Chau Sit? If Haverman didn't show up, and managed to get away, then what? Then he would get away. So what? Let Gilmore worry about it then. Siddler was tired of thinking about the whole thing. His method was to guess, make the intuitive leap. Sometimes it worked. Sometimes it didn't. But it always saved a lot of wear and tear. And it would be safer this way: Siddler had already faced Haverman once and he didn't want to do it again.

A band came into the room, one of the traveling bands from the States that went around Vietnam playing at all the firebases. Some girls were with them, big blonds, big fleshy dollies from Australia. All heads turned.

Watson suddenly rushed in. "Hey Siddler," he whispered harshly, excitedly. "Come on if you want to see something."

Siddler followed him to the broad wooden porch outside the clubroom. The sun had set. From the porch he could look across the entire valley slumbering in the dusk, across Tinh Bien and the Vinh Te Canal and into Cambodia where he saw the dark, brooding hulk of a mountain—the NVA headquarters mountain that Watson had spoken of earlier.

Watson went to a radio table on one side of the porch. Siddler could hear American voices on the radio through a sea of static.

"Listen, listen," said Watson.

"Fifteen seconds and closing," said a crackly voice on the radio.

"Roger, foxfire one, two, three . . ." said a deeper voice.

"What's that?" asked Siddler.

"Those are B-52 pilots, baby," said Watson.

Suddenly the first voice on the radio came through loud and clear: "Hi there you-all in dinkland down there," it said. "From all of us up here, we want to wish you a pleasant evening and . . ." Siddler was looking at the radio in head-nodding puzzlement. ". . . and we just wanted to say to you, GOOD BYEEEEEEEEE . . . YOU . . . MOTHERFUCKERRRRRRRSSSSSS . . ."

Siddler turned his eyes out toward the mountain in time to see some of the first blasts. The Arclight. The force of the explosions seemed to come across the valley in thunderous waves, and Siddler's knees began to shake. He didn't want them to shake, he tried to stop them with an effort of will, but they just kept on, driven by some elemental fear.

"Now that," said Watson, sipping his drink, "that is one *beau*tiful basket of turds." He shook his head in appreciation. Others were hurrying out of the clubroom to watch.

The curved heavens filled with a succession of glittering flashes interspersed with loud booms. The earth shook rhythmically: *wham! wham! wham! wham! wham! wham! wham!* As the exploding bombing pattern walked across the face of the mountain, molten red clouds mushroomed into the darkening sky. Ugly clouds of black smoke drifting across the plain obscured the early moon. Dozens of explosions seemed to come all at once in an earth-shattering cacophony when underground ammunition dumps received direct hits. Shells and rockets from them streaked into the sky at random.

Siddler was mesmerized. Somewhere out there in that devastated world, on that quaking plain, was Chau Sit, where he and Watson would venture tomorrow. Christ, what have I gotten myself into? he wondered.

The crowd on the porch grew. Flashes from the explosions illuminated a gallery of gleaming, drunken faces.

"Ahhhhhhhhhhhhh-n *tar*get!" shouted someone after an ammo dump blew.

Vietnamese troops on the ground in front of the porch chanted, "You die, VC, you die!" They fired bursts from their automatic rifles into the air with staccato chatters.

Watson said "Wow!" from time to time and twirled the ice around in his glass.

Siddler just stared.

Finally he whispered, "Holy Jesus Christ."

10

Siddler insisted on breakfast. Watson wanted to get going, but Siddler insisted. He had a terrible wake-up headache, for which he took four aspirins, and he thought food would help it. Beyond that, he had a theory about a big breakfast when you are going to be out in the field all day and might not get much else to eat: eat it. That's how you keep from fainting in the hot sun, according to Siddler's theory. That's where you get that spare energy that might see you through a tight spot. Fill your belly with food, jam it in. It will make you lethargic for the first hour, but later, when the others are flagging, you will come on strong. You will contend with fear better, because fear is something that has to do with the belly and the belly's reserves. In Saigon it was bloody marys and peanuts for breakfast, but in the field it was good solid food.

Siddler had once known a skinny little Italian war correspondent named Giovanni. Giovanni bragged charmingly about this life style of his, which involved carrying two canteens of red wine and having it for breakfast, lunch, and dinner—sometimes with bread, but never anything else. Giovanni did well on this diet, but then at the end of one long, hot afternoon out in the middle of the delta somewhere he had had a disagreement with the commander of a South Vietnamese tank he was riding and the commander had told Giovanni he could damn well walk home. When little men in black pajamas began chasing him across the fields, shouting at him in French and Vietnamese to stop and have a chat, Giovanni should have easily outdistanced them to the next town—but he didn't. He never made it. He ran out of energy. Fuel. Go power. Just like it says on the Kellogg's box. No shit.

So Siddler had six eggs and a pound of bacon, which made everybody angry in the little American compound because the twenty men who lived there had a certain weekly food allotment and when that ran out it was *nuoc mam* and chicken *a la gook* until the next Saturday. Siddler didn't care. He wolfed it down with hot mugs of sugary, creamy coffee and innumerable glasses of chocolate milk, sopped it up with a dozen pieces of heavily jammed toast and finished off with two sugar-coated doughnuts, all the while returning with an insolent, blank look the stares and glances of resentment and disgust on all sides.

Watson sat beside Siddler and ate a modest breakfast. He spoke of the specific details of the operation, which units would be sweeping where, what to look for. Siddler couldn't understand or remember any of it, so he stopped listening after a while. Detail. He only cared about one thing, and that was the hamlet of Chau Sit. He wanted to see it, and then keep his eyes on it.

"What about that guy who's surveilling Haverman?" he asked as he and Watson made their way in the predawn darkness down to the helipad.

"He's still on him."

"So what's Haverman doing?"

"Nothing. Just working down at Tran's. Spent the day working with Tran's supply officer. Went to Tran's villa after dinner, probably stayed the night."

"Cozy."

"Anyway, he's still around, so you got no worries."

The heavy *whumph! whumph! whumph!* of the rotor blades made Siddler shiver in the semidarkness as they approached the pad. He held his hat. The fragrant, warm air slapped his face with lush splashes. If I could only fuck like that, he thought, like a two-thousand-horsepower aircraft engine . . .

The exhaust pipe on top of Watson's command chopper was sputtering gobs of black smoke in a chugging roar as they walked up. The helicopters all looked big and awkward with their blinking lights. They were pulsating feverishly like giant hearts sitting on skis.

The two doorgunners on Watson's chopper closed the cabin doors for the pilot and co-pilot. Then, zipping up their green flight suits and putting on their heavy combat helmets, the gunners jumped into the small gunnery cubicles on each side and to the rear of the passenger-cargo compartment.

They snapped gleaming belts of ammunition—wicked pointed slugs and long, brass shell casings—into the breeches of the stubby M-60 machine guns mounted on steel struts and sticking out, one from each side of the chopper.

Watson and Siddler stepped up into the chopper, settled into web seats next to boxes of flares and ammunition, and snapped on their seat belts. Watson reached forward to a radio console and picked a set of heavy earphones off a hook there. He put them on, then handed a second set to Siddler, who with some effort was stowing his knapsack under his seat. Siddler put on the earphones and his ears filled with static and voices reading numbers.

The clatter of the blades increased to a deafening roar, making Siddler's whole body vibrate. The chopper suddenly was hovering inches off the ground as the pilot tested his lifting power.

"We're loaded to the tits in explosives," said a voice in Siddler's earphones. "Hope we don't take too much fire until we unload it." When Siddler looked surprised, Watson explained that they could talk to one another over the chopper's internal circuit without being heard by anyone not on the chopper.

They hovered, turning, the whole ship straining, lumbering higher and taxiing to the short dirt runway as if they were on wheels, the other four choppers falling in behind. Suddenly they were lunging forward and up, the chopper's nose tilted slightly down at the still-dark earth.

Looking out through the rushing air, Siddler could see lights in Tinh Bien and, beyond it, the dark, looming mass of the Communist command mountain inside Cambodia. On the intervening plain, he could see the occasional flashes of artillery fire. A flare shot up in the darkness from an outpost somewhere below them and then floated gently down, casting its eerie, pulsating glow over the land.

The choppers were soon high and isolated, the sense of slipping movement over the earth reduced. The top edge of the rising sun hit them. They formed a snakelike line in the sky, a snake of five links wheeling around, bobbing up and down, writhing through the warm dawn, glinting in the sunlight. Siddler felt alone and free, tired and lethargic, drained. He thought of Chi-Chi and the thought began to stir him.

"Tran's bird ain't up yet," said Watson sharply over the internal circuit. Siddler came to with a start. "We'll get a report if Haverman goes up with him or whatever." Watson went back on the general circuit, issuing commands to his other choppers, which began to peel off in different directions; filling himself in on the dispositions of the Vietnamese troops who even now were beginning their slow, sweeping walks through the paddies, hamlets, and treelines below; talking with the tactical air control unit at his headquarters; ordering artillery fire in coordination with the sweeping troops. Siddler fell asleep. His dreams were charges of electricity bolting from his fingertips and through his entire body. Full of yearning and hot desire, his throbbing body cooled by the whipping winds, he dreamed he caressed the smooth, cold black steel of the M-60 machine guns. The gun-smoothness was cool skin, the skin of a hard, bitch-woman. The rifle-woman. The machine-gun-woman. He was firing her, first a shot, then another, then a steady blasting burping stutter from her barrel, his barrel . . .

"Hey wake up goddam it, there it is!"

Siddler jerked awake with a violent rush of blood to his head, sucked in a deep breath, and saw the blazing bright brown of a rice paddy, green trees in a line, and the thatched roofs of a Vietnamese hamlet. Little brown people were moving there, people dressed in drab rags. He felt so close that he thought for a moment that he could reach down and touch them. Then he realized with panic that he was looking almost directly down, that sitting on the edge of the chopper he was held in place only by his seat belt and by centrifugal force as the chopper made tight circles over the ground, turned, it seemed, almost on its side.

"That's Chau Sit," roared Watson.

Siddler drank in the rich detail, still waking up. He didn't know what he had expected: it seemed like any other Vietnamese hamlet to him. There were other hamlets in nearby treelines. Together, the hamlets of a few hundred persons each would form a village. Little figures were already working in the golden fields all around, some of them walking behind oxen. There was a mountain in the distance, bluegray and hazy in the harsh morning sunlight. In the sky around them other helicopters hovered and moved, some close and others distant specks. The earth seemed flat and easy to understand, but the world he was in seemed all swirling air, movement through space, swooping confusion.

"Tran's up," came Watson's voice over the radio. "Haverman's with him."

Siddler looked at Watson, sitting next to him, intent on what was on the other side of the chopper, almost sitting up out of his webbed seat, straining with the effort, the veins standing out in his neck.

Watson glanced at Siddler and smiled boyishly. "Looks like you win, pal," he said. "Those bastards have been in this area ever since they came up. He's already tried twice to order me over to the other side of the mountains."

Siddler was fully awake now.

"Where are they?"

"Over there," said Watson, "the one with the red nose." He pointed into space on his side of the chopper, and told the pilot, "Jim, see if you can get a little nearer Tran so Mr. Siddler can get a look at them."

"Yes sir," said a collegiate voice.

They wheeled through the sky, shooting down toward the earth, leveling off just above the rice paddies, and skimming along at a tremendous rate of speed. It was like a widescreen Cinerama film Siddler had once seen—the total sweep of forward motion, racing through worlds and colors and atmospheres, sweeping along just over the heads of startled peasants, cringing slightly, holding their conical hats to keep them from being blown off by the passing blast. Siddler thrilled to the sense of speed and power. So this is war, he thought. Jesus, give me more of it. It's no wonder we fight if it's this much fun. If I could fly around in machines like this, I'd become a general. It was like going a hundred and fifty miles an hour in a car, but with absolutely no sense of danger. It was like being a bird. It was like a dream.

They shot under Tran's chopper and then up around it, away from it again at a distance. It was a dark green, curved chunk hanging in the sky, its red nose bright, its passengers tiny specks with no features. It was much higher than they were, not moving as fast. They could not understand anything on the Vietnamese radio circuit that Tran used most of the time. So close to Tran's chopper, yet they might as well have been in a separate world.

"He's high so he can watch the whole operation," said Watson. "Look down there."

Siddler did and spied a long green line of figurines— battalions of troops walking slowly, advancing across great wide rice paddies.

"Christ," he said, "you can order them around from up here like puppets."

"I can't," corrected Watson. "Tran can. But I sure can put them in a lot of deep shit depending on what I decide to do with the airpower. That's my baby, just take a look up there."

Siddler saw silver streaks in the sky above where Watson was pointing. Jet dive-bombers.

"Spotter planes control them," said Watson, "but I've got the final say on where and how long and when they bomb, and Tran knows it."

They flew back toward Chau Sit and in a few minutes

were circling over the little hamlet again. Watson occupied himself on the radio, then said, "I gotta go refuel."

"Shit," said Siddler.

"No choice," said Watson, "unless you got high-octane spit."

Suddenly the deafening clatter of machine gun fire splattered into Siddler's ears as the doorgunners opened up with their M-60s. A stream of burning-hot empty casings played against his head and shoulders and clattered on the metal floor and bracings of the cabin.

"Tracers!" shouted Watson. "We're taking fire from that fucking hamlet! Get outa here!"

The chopper dived abruptly, gaining speed rapidly, and churned away from the hamlet. The clatter of the machine guns continued. Siddler saw one of the doorgunners behind him standing up and straining forward against his chattering weapon, aiming it straight down, his arms shaking as if he were operating a jackhammer, held in place only by his seat belt against the violent motions of the chopper. Then the firing stopped and the gunner sat down, slumped in his seat, grasping his right side with both hands, an agonized expression on his young, surprised American face.

Siddler didn't look back again until they were about to land at the American compound to refuel. When he did, the gunner's head was bobbing with the motion of the chopper, hanging out over space.

"He's dead," said Watson.

A jeep came and medics took the dead gunner away on a stretcher while the chopper was refueled.

Watson's face was tense and angry.

Siddler poured a G-and-T for himself, drank it quickly, but offered nothing to Watson.

When they were airborne again Siddler noticed that they were not alone. Three lean helicopter gunships were in a fan pattern slightly above and behind them.

"Cobras," said Watson. "We got enough firepower with us to wipe out half of Saigon."

He spoke into his network radio: "Control to Rangers seven, eleven, and five."

Siddler heard three answers: "Come in, control."

"We're going to make passes at Chau Sit," said Watson. "We take any fire and I want you to open up."

All three: "Roger out."

They came toward Chau Sit only a few hundred feet off the ground, the Cobras staying well behind. Siddler didn't like it much. He had come to surveil, come to do a little narc work for some 'crat in Washington named Gilmore, and now all of a sudden he was getting involved in the actual fighting of a war, flying like an insane man low over an enemy position to draw fire. A guinea pig, or something worse. Siddler didn't like it at all, didn't figure this came under the category of duties that he could legitimately be expected to perform, was having trouble keeping his knees under control again, and found that he was instinctively pressing them close together on the notion that if they took fire again at least the bullets would have to get through his legs before reaching his balls.

He was about to voice his concerns when he saw a helicopter sitting on the ground almost directly below them. It was on the edge of the hamlet and had been hidden by trees until they were directly over it. The chopper had a red nose—Tran's chopper.

"Holy shit," said Siddler. "There they are."

"He's shot down," said the pilot.

"No, he's not," said Watson.

Soldiers were running to and from the helicopter, carrying boxes each way. The chopper's blades were moving around at idling speed. Siddler could see several men standing near the chopper.

In a few moments, they were beyond the hamlet and the chopper on the ground was out of sight.

"Go higher and circle," commanded Watson in a shout. "Come in Rangers seven, eleven, and five."

"Roger," said the three voices from the Cobras.

"Back off with those Cobras," shouted Watson. "Go high and circle, wait for my instructions. Do not, repeat *not,* fire at anything under any circumstances until you have **my** order. Over."

"Roger, control, understand hold fire, out."

"Christ," broke in Siddler with a roar, "let's go in and get 'em. Look at all the scag they had—they must have fifty kilos or more."

Watson, who had succumbed to Siddler's arguments in the command bunker, did not succumb to them in the field. This was his province: battle, tactics. It was this for which he was trained, for which his instincts were honed, and those instincts told him as surely when to hold back as when to go in shooting.

"You keep outa this, Siddler," he shouted. "That's all I need—to wipe out the goddam Vietnamese commanding general and create an international incident."

"Goddam it," shouted Siddler, "we gotta go down and arrest those bastards."

"Listen you idiot, for all we know that's powdered milk they got down there. Now shut up and let me handle this."

Siddler turned away. Judging from the number of boxes that were going back and forth, the shipment was a huge one. Siddler realized that Watson was right, because Haverman and Tran held the controlling cards in this situation. The best Siddler could do was watch and follow.

Suddenly Watson was moving, standing, shoving at the boxes inside the chopper.

"Let's dump this stuff," he shouted. "It's ammo and if they hit it we'll go sky high."

Siddler let him work. He wasn't going to unfasten his seat belt and stand up in a moving helicopter with wide-open sides that you could fall out of.

The heavy boxes went overboard. Watson sat down again.

"All right, Jim," he said. "Go down. We're gona land beside them."

Siddler stiffened. They circled down.

Tracers spewed up at them from the ground.

"Up, up, up!" shouted Watson. They swerved to avoid the tracers and raced away from the hamlet. When they were a mile away, the pilot began circling back toward the hamlet and gaining altitude again.

As Siddler looked out his side, he saw Tran's chopper rising into the air and going in the other direction.

"They're heading toward Saigon," said Watson. "Follow them."

"Yes sir," said the pilot.

Siddler looked back a few moments later and saw the Cobras diving straight toward the hamlet, dipping fast with rockets spewing from their pods. He saw the fiery explosions and huge balls of angry black smoke blossoming from the earth as the napalm hit.

Tran's chopper was a tiny dot over the vast plain. After a ten-minute chase, the distance between them remained the same.

"Shit," said Watson. He looked at Siddler and shrugged.

"No sweat. They'll never get the shipment on the ground in Saigon before we swoop in on them."

Siddler thrust his head out as far as he dared, straining to look ahead. Over the haze of the plain he saw Tran's chopper, a pale green blob hanging in the air more than a mile ahead. The 120-knot wind made his eyes water, and he drew his head inside.

"I don't think they'll land in Saigon," he shouted.

"Why not?"

"They'll land at a Vietnamese firebase somewhere and blow us away when we come in after them."

"You should have been a goddam tactician."

The pilot broke in urgently.

"Hey," he said. "They've pulled up."

Siddler and Watson could see immediately that Tran's chopper was getting closer, growing larger. They saw the red nose.

"Jesus," shouted the pilot, "he's coming at us."

"Hey machine gunners," shouted Watson, "get those machine guns warmed up."

The new man on Siddler's side limbered up his weapon and chambered the first round.

When the choppers were close, Siddler saw that Tran's was much higher.

"Don't let him top us!" shouted Watson. "We can't shoot up through our blades!"

"Too late!" yelled the pilot.

Siddler pulled out his .45, ready to fire.

The two helicopters maneuvered for position, each going higher and higher. Siddler only got glimpses of Tran's chopper from time to time: the tense faces of Tran's pilots, the passenger compartment where Tran and Haverman must be—but he couldn't see them.

"We're taking fire" came a near-scream over the earphones from the co-pilot. "Christ . . . we're hit . . . Jim, Jim, Jim . . . Christ, Jim . . ."

A gleaming red stream of tracers zipped by, moving toward Siddler. He ducked and at the same instant felt the chopper plunge. The chopper leveled again after a few seconds as the co-pilot brought it under control.

Suddenly the door gunner opened fire, and a stream of hot shell casings played over Siddler, spewing into the cabin. Tran's chopper appeared beside them, so close Siddler could make out the rivets in the skin. Siddler found himself looking into the same pair of intense blue eyes, the same reddish face that he had seen two days before—Haverman.

The split-second scene was frozen in time: the heavy lumbering chopper beside them, the profile of the Vietnamese pilot, the detail of his helmet down to the Vietnamese lettering in white, his intense concentration, the Vietnamese door gunner at the rear, his gun's deadly little hole seemingly aimed right in Siddler's face and spurting fire, Haverman braced against the cabin with both arms and looking directly into Siddler's face, and farther back in the dark cabin a shadowy face: Tran.

Then the instant shattered, and Siddler felt only the dizzying drop of his own chopper, heard a scream from the pilots' compartment, and realized blindly that their engine had stopped. The chopper fell sickeningly, its big blades no longer under power but flapping in the wind, barely keeping the craft from dropping like a stone as it plunged out of the sky.

The chopper hit hard, rolled over, and came to a rest against a dike dividing two rice paddies.

Siddler rolled out into the dust—he didn't know how his

seat belt got unfastened—and started running as hard as he could along the top of the dike. He heard screams behind him and, glancing back, saw Watson pinned under the chopper.

"Christ . . . Siddler . . . my leggggsss . . ."

The anguished call drifted across the still, hot air. Siddler hesitated for an instant in mid-stride but then kept running as fast as he could toward a treeline ahead. The choice was instinctive, final: one he had made a dozen times before. He bore the dozen scars on his soul, but he bore them well.

"Sid . . . luuurrrrrrrrrrrr . . ."

In the treeline he flopped to the ground and looked back across the field. He panted hard and waited, his mind strangely composed. He realized that he held his .45 grasped tightly in his hand.

He heard a helicopter. It got closer and closer. He couldn't see it.

Suddenly the noise increased to a roar and Tran's chopper raced across his field of vision, diving for the downed chopper with machine guns blazing. Two rockets streaked down. As Tran swerved away and up, the helicopter on the ground exploded, sending a huge red ball of fire mushrooming into the sky.

"Jesus," said Siddler, shifting uncomfortably on the dirt.

Then out of the corners of his eyes he saw movements, and when he looked directly at them he could see tiny figures, dressed in black and carrying what looked like sticks or rifles, running toward him across the rice paddies. He could hear their shouts dimly, high-pitched cries in Vietnamese carrying far across the hot air, punctuating the crackling roar of the burning chopper.

Christ, he thought, those ain't friendlies.

PART III

Closing In

11

Gilmore put his right eye to a twenty-power monocular that sat on a steel tripod in a room like the inside of a packing box. The view zoomed across a wide street, across a partially graded parking lot on the other side, and stopped at a rambling brick house with a four-car garage and a long black awning down to the street from the front door.

The awning said: *McVey's Funeral Home.* Gilmore shifted the view. The house sat back from the main road behind a chain link fence. It was quiet in the suburban morning.

"Big Nick arrived there last night about ten from his Irving Street apartment," a man said to Gilmore. "He's been in there ever since. You swing that spy glass you'll see down the road there a utility truck working a manhole. That's the bug team setting up. We've put a tap on McVey's phone, too."

The man's name was Everson. He was a plainclothes black detective on loan from the metropolitan police department. He was a member of the department's "old clothes" division and was dressed like a dimestore cowboy in green bell bottoms, a kelly green shirt, wrap-around sunglasses, and white leather boots. He wore a .38 caliber Smith and Wesson on his right hip.

A tape recorder sat on a card table in one corner of the filthy room. A wraithlike black man wearing earphones sat next to it, playing solitaire, ignoring Gilmore. A Remington 12-gauge shotgun leaned against the wall next to him.

"McVey is John McVey," Everson said. "He's Nick Westley's uncle and has run that funeral home for years. One of the first blacks to move out of the District after the war. It was all white here then. Now it's almost all black. Prince George's County cops say McVey is clean so far as they know."

"But Nick came here after the Guardhouse raid last night," Gilmore said.

"They left the apartment on Irving Street at nine-oh-five," Everson said, leafing through a notebook. "At first we thought they'd been tipped, but then I remembered—they must have heard it on the news. The cops put out a statement in time for the nine P.M. slot."

"What kind of calls have come into McVey's?"

"Just funeral stuff. It's the only black home out here, and they do a heavy business."

"Let's see the log."

"We didn't get set up until a little after midnight," Everson said, "so we missed a lot."

Gilmore looked at the entries in the wiretap book. The brief notations indicated routine business. "Nick hasn't called out on this phone?"

"Not since we been listening. What's he going to say to anyone—that he lost three men in a shootout with a narc? You don't boast about that."

"Sure. But you don't go to your uncle's funeral home after a disaster like that, either. You split completely."

"For what it's worth, he thinks he shook whoever was following him. He thinks he's home free." Everson grinned. "As a matter of fact, Nick *did* shake the tail."

"How'd you find him?"

"We never lost him," Everson said pointing to the ceiling of the room. "Had a helicopter from traffic do it."

"Nice."

"Here's a call," the man in the corner said. He flipped a dial on the recorder and they listened in silence as a woman, her voice breaking, called to seek advice about an impending funeral. A heavy, respectful voice responded. The conversation went on for a few minutes and then ended.

"That's McVey," Everson said. "We haven't seen enough to know exactly how the setup works. I think there's McVey, his wife, one or two assistants, and some day help to drive the limo, keep the hearse polished, stuff like that."

"I suppose we could go in there and yank Nick out," Gilmore said. "If we do that, what do we net?"

"Nick on a conspiracy charge. Conspiracy to assault a law enforcement officer."

"Never be able to prove it. Even with the taps and bug from the Irving Street apartment there's not enough evidence. They never talked about it, did they?"

"All I ever heard was some half-baked talk about offing the pig.' "

"That was me."

"Yeah. As it turned out, it was."

"All right, keep me posted," Gilmore said. "I'll be at the task force."

"Sure," said Everson. "Anything unusual."

Holt was sitting comfortably in his chair. He looked rested and cheerful. Gilmore told him about Nick. Holt nodded and smiled. "That's fast work. I'll assign round-the-clock coverage."

"Good."

"Army Criminal Investigations is onto something."

"Already?" Gilmore was surprised.

"I think this stuff has been there all along, Dan, we just haven't looked in the right places. Now we're starting to look and it's under our noses." He read from some notes.

"We gave them the orders we found in Roulette's apartment at eight this morning. It's now eleven-thirty. In that short time they've turned up a guy named Cronder, who looks like a possible suspect. He's a warrant officer, oversees the paperwork in the adjutant's office at Fort Myer. A career man. Two marriages. Two divorces, five kids. Unhappy man in recent months, his associates say.

"Cronder's in a perfect position to engineer a forgery like this. He handles mountains of papers that flood in there. He sifts it in, he sifts it out. He *knows*. He has access to everything. The handwriting types have found that the adjutant's signature on those orders is genuine. It wasn't put there by someone acting in his behalf. That gives it a nice authentic touch, if you collect autographs. But the adjutant is red-faced. He doesn't know what he signed."

"They never do," Gilmore said.

"The adjutant thinks now that his faithful servant Cronder has been putting one over on him."

"What's in it for Cronder?"

"Maybe he needs to do people favors. He's a gambler."

"Horses?"

"Cards."

"Roulette Parkes is a playmate?"

"Could be," Holt said.

"And our friend Cronder the warrant officer is hip-deep in IOUs," Gilmore guessed.

"That's what his office chums think, with a little prodding from Army Criminal Investigations Division. They say he's been having tight jaws lately. He's losing heavily, these people think."

"It has to be," Gilmore said. "He plays, he loses, he owes them. The debt piles up. They'll forgive the debt if he does them one small favor. He agrees. He plays some more, they whip his ass with their sleight-of-hand specialist. They threaten, he bends, they get their fake orders."

"Probably something like that," Holt said.

"It's Nick's kind of angle," said Gilmore. "You do them a favor or they bust your head. And then your kids' heads and your old lady's. Wherever you live, they come and call."

"The Army is still working it. They'll tap his phone, surveil him. They won't move to apprehend until they get a go-ahead from us. That's the arrangement for now."

"Sounds fine."

"Now for Part Two." Holt reached into his desk, pulled out a wad of paper, and flipped it on Gilmore's desk.

"You read that rocket, Dan," Holt said, "then ask yourself what you did for a living this morning."

Somehow the telegram from Siddler had been crumpled in transit. Now it looked as Gilmore imagined Siddler himself to look: rumpled and dirty. He read.

It rambled. It swore. It went up blind alleys. It backtracked. It grumbled, groveled, insulted, lied. It weighed a ton.

But at the same time, it sketched. Suggested. Told an anecdote. Described. Showed a pattern. Informed.

"Boil out the horseshit and the whole message might cover half a page," Gilmore grumbled.

"You don't boil horseshit out of Siddler. That comes with the model, just like wheels come with a car."

Gilmore's mind reached out and he could see Siddler, a

fat, overage man scribbling his message, his brain in a frenzy of activity, his pride sticking out like the prow of a ship, the pride and the need for some voiced approval hand-in-hand, escorting the information like precious cargo.

Gilmore marveled at how this slovenly message, with its swear words and asides, it complete rejection of brevity and of cool form, had been coded, tapepunched, fed through a machine, converted to signals, transferred at.the speed of light, then gathered in again, amplified, decoded, taped in plain language, and spewed out of the flying head of an IBM Selectric, collated, and transferred to his office. It seemed Siddler, recreated before his eyes.

URGENT//SUPERSECRET AND CONFIDENTIAL
PRO CUSTOMS CONUS//DANIEL GILMORE TASK FORCE WASHINGTON
DE CUSTOMS SAIGON//RALPH SIDDLER//US EMBASSY SAIGON
INFO ADEE: NONE RPT NONE

HELLO GILMORE. THE WHISTLE IS RONALD E WHISTLER SPECIALIST FOUR US ARMY FREQUENT FACE IN SAIGON. SMALL TIME GAMBLER KNOWN DRUG FRINGE HANGERON. WHISTLER MUST BE A COURIER LADEN WITH HEROIN AND RETURNING WITH IT TO CONUS PROBABLY TO ARRIVE DOVER AIR FORCE BASE TWENTY JUNE AS PER YR MSG VIA DIRECT MILITARY CHARTER FLIGHT FROM SAIGON. FLIGHT INFORMA-TION TO BE SENT SOONEST TO YOU FROM MY ASSOCIATE, ASSWIPE NAME OF TRAGER.
BOGUS ORDERS DECODED AND INDICATE THAT FAKE SERVICE-MAN PAUL FLYNN JOHNSON NE ROULETTE PARKES DETAILED TO HEADQUARTERS UNIT FIRST LOGISTICAL COMMAND MACV SAIGON PLACE CALLED SHED SIXTEEN. NO MEANING TO YOU BUT THE APPARENT BRAIN OF MASSIVE FIRST LOG WHICH CONTROLS EVERYTHING FROM AMMO FOR HUEY COBRA TO ZIT REMOVERS FOR PUNK GRUNTS.

SENIOR OFFICER PRESENT AT SHED SIXTEEN IS ONE RUPERT
KAISER HAVERMAN ARMY COLONEL REDHAIRED SONOFABITCH
BY LOOKS MAYBE LOONY AFTER TWO WARZONE TOURS OVER-
DOSES OF ARMY LIFE AND OF YOU KNOW WHAT. EXCELLENT
WAR RECORD BUT HAD ODDBALL TASTE FOR YOUNG GIRLS
MAY EVEN HAVE KILLED YOUNG PERSON KNOWN TO DEAR
FRIEND. MEANING AGAIN NOTHING TO YOU PAL BUT MUCH
TO ME AND I'LL KILL HIM IF I GET THE CHANCE. WILL
KEEP CLEAR UNTIL HAVE SAFE AMBUSH. HAVERMAN OWNS
BIG HONG KONG DEPOSITS UNKNOWN TO GENERAL PUBLIC
SAYS FRIENDLY CHICOM BANKER WHO SHOULD KNOW.

LOCAL MINISTER OF CORRUPTION HINTS BUT OFFERS NO
PROOF OF HEROIN CONNECTION BETWEEN HAVERMAN AND
LOCAL SLOPE GENERAL NAME OF TRAN. AGAIN NO MEAN-
ING TO YOU BUT A BIGBORE PAIN IN THE ASS TYPE OF GUY
OUT HERE.

NOW CLEARLY ENOUGH SHADY INFORMATION KNOWN TO ME
ABOUT HAVERMAN TO SEEK FULL ASSETS AUDIT OTHER IN-
QUIRIES LEADING TO COURT MARTIAL AND SO FORTH. BUT
WILL HOLD TONGUE AGAINST BETTER JUDGMENT I'M SURE
OF SMART-ASS WASHINGTON AND LOOK INTO THESE FURTHER
COINCIDENCES AS FOLLOWS.

HAVERMAN OPERATION CONTROLS MORGUE AT TANSONNHUT
AIRPORT WHENCE FRIEND LUCKETT SHIPPED OUT UNDER
MISTAKEN IDENT OF JOE DIMALCO. INCREDIBLE DETAIL
AVAILABLE AS TO HOW THEY PREP RETURNING SERVICEMEN
AT MORGUE. RICH LOAD THERE BUT NO TIME TO BORE YOU
NOW. WILL SEND MORE LATER MAYBE BUT THOUGHT YOU
SHOULD KNOW AT LEAST OF POSSIBLE INQUIRIES LATER.
MAYBE LUCKETT SAW INSIDE OF MORGUE AND THAT WAS
THE END OF HIM MAYBE HE DIDN'T SEE IT. WILL SEEK
MORE THERE LATER.

MEANWHILE WILL HEAD SOUTH TO SEVEN MOUNTAINS HOME
OF NVA WHERE HOPE TO CHECK LUCKETT'S MOVEMENTS
PRIOR HIS UNTIMELY DEATH. WILL FILL IN ON MORGUE,

PRECISE AMOUNT OF HAVERMAN GOLD IN HONG KONG BANK, OTHER ALLEGATIONS UPON RETURN. IT WILL KEEP BELIEVE ME. THIS UPCOMING TRIP MAY BAG ONE HORSESHIT COLONEL ONE DINK GENERAL ONE SPIC EMBALMER ONE SOMETHING ELSE MAYBE KITCHEN SINK FULL OF SCAG. MESSIEST GOD-DAM ASSIGNMENT SINCE KOREAN WAR RETRIEVAL OF THREE CHINAT GINK AGENTS BY SAMPAN FROM UNDISCLOSED TEEM-ING MAINLAND HARBOR JAWS OF RED DRAGON MENACE. DEMAND RAISE. LIVING STYLE SUFFERING FROM AMERICAN-TRIGGERED INFLATION. HELP DEMANDED. MORE TO COME. SIDDLER.

"I told you the sonofabitch would come through," said Holt when Gilmore had finished.

Gilmore nodded. He wondered where Siddler might be, hoped he was sitting in some plush barchair with a frosty cool one in front of him, savoring his moment, knowing that his cable was making things clear and giving them vital connections.

Gilmore slept well and spent the next morning completing his report on the Guardhouse shooting of two nights before. He cleaned up some other paperwork and then got a call from Everson.

"I've been going through the tapes from last night and I think you ought to hear this one call," the man said.

"Play it."

There was a pause and then Gilmore heard the tinny replay of a telephone ringing. It rang three times and then a voice said, "Hello?"

"Andrew M. Samuels," said a quiet, nervous voice. "E-3. PFC. Five two seven, six five, oh seven four three."

"Thank you very much." The conversation ended.

Everson came back on the line.

"So what?" Gilmore asked.

"You ever hear a conversation like that before?"

"No. But I'm not a mortician."

"That call came in at ten in the evening. We've been working this morning to find out about McVey. There's a mortician's licensing board that approves these guys and makes available to the services a list of qualified, recommended funeral homes to handle military burials. McVey is on that list and he has a contract with the Department of Defense to handle some of the funerals of war dead in this part of Prince George's. It's a contract that most morticians want. It more or less guarantees them an income."

"So they get these kinds of calls."

"No. Normally they are notified by message by the Department of the Army, or whatever service the person happens to be from. We've been listening now for almost two days to this guy's phone and that message sticks out like a sore thumb. He never got anything like it before and there's been nothing like it since."

"Keep listening. I'll have the Pentagon start checks to find out about Samuels—who he is and what he does."

Within three hours, a Department of the Army spokesman had the information and read it back to Gilmore. "Samuels, Andrew M., with the same serial number you gave us, Mr. Gilmore, was a member of the One Hundred First Airborne. He enlisted last August from his home in Landover, Maryland, attended jump school at Fort Bragg, and shipped to Vietnam 19 March of this year. He was killed in action five days ago."

"Has his body been shipped home?" Gilmore asked. His pulse jumped in his head.

"It's en route for a burial by a civilian funeral home per standing DOD instructions."

"Do you have a date of arrival?"

"Yes, sir. June 20."

"Is the Samuel youth . . . black or white?"

"Black."

"What funeral home is handling the burial?"

"McVey's, off Pennsylvania Avenue Extended, just beyond the District line."

"Is there anything else you can tell me about the arrival or the arrangements for burial?"

"It says here in this message from the First Log Command—that's the originating authority for this information—that the remains are not suitable for viewing." The man paused. "Can I ask what this is about?"

"You can ask. I can't say."

"Oh."

Gilmore put the phone down and with building excitement went to see Holt. Whistler was coming on June 20, the same day of arrival as the body of Samuels. Whistler's arrival date had been known to Roulette Parkes and, by extension, to Nick. Nick's Uncle McVey got a message telling him of Samuels's arrival. Same day. Same suspect.

As Gilmore and Holt talked, Gilmore found himself astounded at the pattern that was beginning to emerge. And astonished to realize that the information break came from the pigeon, who in turn had surfaced because of publicity over the shooting of Roland Drinkwater, a chance encounter, isolated in time and memory, from which new events now were flowing mysteriously and powerfully toward them.

Holt answered a telephone call, made curt inquiries, scribbled some notes, and hung up. "Warrant Officer Cronder made a call last night from his apartment on Route One south of Alexandria. He called McVey's funeral parlor at ten P.M. He's the fellow you heard telling about that Samuels kid. Isn't that a pleasant coincidence?"

"Indeed. Will the Army do us another favor?"

"We can ask."

"Haul Cronder's ass in where we can talk to him."

"I thought you might be interested in that. The arrangements have been made. By the time you and I arrive at the Fort Belvoir stockade over in Virginia, some fifty minutes by non-rush-hour traffic from here, the warrant officer will be gathered up and deposited in a small, empty room they keep for just such occasions. I thought taking him to Fort Belvoir would be good, since it's off the beaten path and who knows what kind of security there is at Fort Myer? We can have Cronder all to ourselves. Just us and CID."

Gilmore peered through the small window in the door. A thin, lanky man with slicked-down hair was sitting at a table, staring blankly at a dun-colored wall. He was slowly working on a sandwich. Steam rose from a coffee cup. Smoke from a cigarette curled at the edge of the table.

"We gave him a bite to eat because his goddam stomach was growling so," said an Army captain from CID. "We've softened him up a little and he's sweating. You bust in there with us and we'll get it all in one big dump. Hookay?"

Gilmore nodded. They went in. Cronder whirled around as they entered. "I ain't done nuthin'," he protested. "I demand my rights. What's this all about?"

"Shut up, Cronder," said the CID captain wearily. They ringed themselves around him. The captain, whose name was Jones, tipped the cigarette on the floor and stomped it out. He lounged next to the warrant officer. "Now these men here are from the Justice Department and they are going to throw your ass behind bars for maybe twenty, thirty years if you don't quit lying."

"I'm not lying," Cronder said. "I'm not covering up. I'm telling it true."

Gilmore studied him. "How much do you owe, Officer?"

"Let him tell it to the judge," said Jones.

"Now wait a minute, Jonesy," said Gilmore. "Man has pressures, man gets into a hole, he has to help himself. I can understand that even if you can't. The officer here is going to tell me how he got in such a hole."

Cronder swallowed and said nothing.

"I've been there myself one or two times, I can understand it," Gilmore continued. "You get in a hole, you get behind on a payment or two, and then they're after you and they won't leave you alone. It's just a case of a fella getting a little behind in his payments. Isn't that right, Officer?"

Cronder hesitated and then gave a twitch of a nod.

"Now, since you're leveling with me, I'm going to level with you," Gilmore said, smiling warmly. "I'm not what Jonesy there said I was. I'm not from the Justice Department. I'm nothing more than a Customs inspector."

"Christ," Jones broke in. "You think he's going to believe that?"

Gilmore swept the badge into view. "So there's no question of my being a lawyer who's going to send you to jail. Nothing like that. But I *do* have a problem that you can help me with. How about it?"

Cronder said nothing.

"Like how come a smart warrant officer like yourself would be calling a funeral director in the middle of the night with the name of a dead serviceman?"

"I didn't do it."

"Sure you did. We got the tapes."

"I didn't," Cronder said.

"Yeah," said Jones. "You lying bastard, sure you did."

"Now wait a minute, Jonesy," said Gilmore. "Maybe he didn't. You probably didn't know who you were calling, did you?"

Cronder shook his head.

"That's what I thought. The reason I thought that is because it really wasn't much of a call, just the name and serial number of some dead soldier. Right, Officer?"

"Yeah," said Cronder. "Yeah, what harm was there in that?"

"How much was it worth?"

"Oh, Christ I don't know," Cronder said. "I don't know. It got so high after a while I couldn't keep track. Maybe six, seven thousand in recent months. I just don't know." His shoulders slumped. "When I'd make a call like that, it would be worth maybe five, six hundred offa the debt."

"You're a gambler, aren't you, Officer?" Gilmore asked.

"Now and then."

"You lost heavy, and then you found a way to erase the debt isn't that how it worked?"

"Yeah . . ." There was silence. Cronder's body heaved with a huge shrug. "I used to be good at cards. I learned early. I played a lotta cards in my time. But I never met no one like that kid. Roulette Parkes, Jesus, he's a good card-shark that kid. I knew he was cheating me after a while, but you get into those things, you get stubborn, and then you get scared. I got plenty scared. And then I cut a set of orders for them." He glanced around for support.

"I owed them quite a bit, several thousand, and one day

I got a call at the office, if I do such and such they'll forget a thousand bucks and who's to be the wiser? Now, you gotta understand the military supply system, gotta know how it works. It's nuthin' to do what I did. Guys are doing it all the time. You remember that sergeant major—the guy who was into the PX in Saigon? That's only the tip of the iceberg." He paused.

"Shoot, there's guys I know been cutting orders for themselves for so long they could do it in their sleep. It's just a big airline, that's all it is, and you can order yourself anywhere you want around the world once you get the hang of it. So I cut some orders for some fellas? So what? It ain't no federal case. Is it?"

"No," said Gilmore. "Not yet."

"I mean, I was the guy in the middle. I'd get a call at work. Long distance."

"From New York? Los Angeles?"

"Saigon."

"That's easy," Jones said. "The military telephone system is probably the best in the world. If you know the right numbers, you can call instantly anywhere on the globe. Anywhere."

Cronder nodded.

"You're sure it was from Saigon?" Gilmore asked.

"No. I just assumed it was."

"Why?"

"Because the guy would say such and such a name and serial number. The first few times it happened, I did like I was told, passed it along to the next place. But then I got nosy. So *I* did some checking and discovered that I was passing along the names of dead men. It didn't bother me much after that. They were guys who'd been killed in the war. And sometimes I'd cut a set of leave orders for some guy. That's all I done. I got two ex-wives and a buncha kids to worry about."

The next day, Trager's telegram arrived bearing the flight number for Whistler. It matched the flight for the Samuels youth. It was a flight direct from Saigon to Dover Air Force Base, Delaware, via Honolulu and Travis Air Force Base in California. Gilmore and Holt worried about the problems of surveillance, about not losing Whistler on the huge air base, and then having to trail him until a connection was made.

"There must be a shortcut," Gilmore said. "Couldn't we just seize Whistler and his baggage and the Samuels coffin?"

"Sure. But if we do it at Dover, where Whistler is likely to be met by his own people, then we're blown. They'll know we've penetrated almost all the way."

"What if we shortstop him?"

"How?"

"Divert the plane. Scramble aboard, get Whistler off, and take our time going through his stuff. Then we don't have to trail him anywhere. When we get him with the scag, we've got a courier by the nuts."

"Sounds good," Holt said. "But which airport?"

"Andrews," said Gilmore. "We're going to pluck him off that plane before he can make a squeak."

"Andrews? That's the President's air base, Dan. They aren't going to like pulling a stunt there."

"Well, unless the President himself gets caught in the middle, it isn't going to amount to much. The plane is nearby and it isn't a shrine. It's an air base with real Air Force planes and a runway you can land jets on. It also happens to be about thirty minutes from here, which gives us flexibility and convenience. And those are things we want."

"But if we pull a raid and get both Whistler and some heroin what have we done? We haven't gotten Nick. We haven't gotten Haverman. We come in about halfway between."

"Whistler knows both ends. He gets his orders to Saigon from Nick. He gets his travel orders back to the States from Haverman. If we get him escorting, say, fifty pounds of heroin, how many years could he get put away for? Twenty? Fifteen? Long enough so that when he gets out, he won't be a kid anymore. He'll know that and if he doesn't we'll make sure he learns it. When he does, he'll sing for us."

"What if Siddler and Trager are wrong—there's no heroin?"

"All right, it's weak there. But now's the time to fish or cut bait—either Siddler's as good as you say or he isn't. If he's wrong, he's through . . ."

"And so are we."

"The biggest problem will be getting Walker to go along with it. I'm going to need some sort of White House clearance to open the base up to us. He'll play very hard to please over the whole thing." Gilmore shook his head.

12

Andrews Air Force Base was recovering from the heat of the day when Gilmore and Holt arrived. Several light observer aircraft practiced landings and takeoffs along the outer ends of the main north-south runway, their tiny engines snarling faintly.

It was seven o'clock on June 20 and the sun slanted low across the rolling Maryland countryside around the air base. The low buildings, clusters of hangars and workshops and the quarters for enlisted men and officers, looked peaceful.

Parked back in a cordoned, guarded area was Air Force One, regal in its white, blue, and gold trim and silvery undersides. Next to the President's plane stood Air Force Two, identical in every detail, and then several rows of smaller jet transports in a variety of rich colors, from the white-with-red-bands of the Coast Guard to a highly polished silver Air Force jet. It was the royalty of the Air Command. There was a blending of color, form, and function in the aircraft that somehow set them apart from the rest of the base.

"That's what it's all about," said the duty base commander, a tall, leathery colonel. "That's presidential stuff, Mr. Gilmore. The finest aircraft ever operated by any nation in the world."

Gilmore nodded. "I don't believe this will take too long, Colonel. Ten minutes at the most. Then they can be on their way."

"I hope you're right, Mr. Gilmore," said the Colonel, his faded brown eyes darting uneasily. "This is presidential country out here. We do things by the book."

"I know this isn't by the book, Colonel," Gilmore said quickly. "That's why the White House itself approved it. The President's aides know what we are doing."

"Well, goddam it, that's true," the colonel said. "But like I said to my aide, what in hell do those civilians *really* know, anyways? They don't know when the President might have to come out here and take off, or his foreign affairs man, or somebody like that. And what would they find here? The goddam base filled with a buncha narcs." He rubbed his chin, the fingers scratching against the day's stubble. "Well, I guess I filed my views with that man Walker. He knows where I stand."

"He told me of your concern." Gilmore adopted a formality he hoped would sound sufficiently military to get this tall yuck back in his office.

A solemn-faced air police sergeant stuck his head out of the door of a jeep parked a few yards away. "Tower's calling on the radio," he said to Gilmore. "They say flight nine-eight-two has just turned on final, Mr. Gilmore."

"Okay." Gilmore turned to the colonel. "Thank you, Colonel. It's been very good of you. I won't forget this, and you can be sure the White House will hear of your cooperation."

"Any time," the colonel said, looking past Gilmore at the light planes that were bouncing across the field.

Gilmore climbed into the jeep and settled into the right-hand front seat. "Let's go."

The vehicle burped to life and the sergeant rammed it into gear. Robert Holt sat behind him and next to him sat a black air policeman who packed a long billy club, a .45 in a well-oiled holster, and a .16 gauge automatic shotgun.

"Nine-eight-two has signaled wheels down," the tower yapped over the radio. The jeep picked up speed, bucking down the taxiway toward the far end of the main runway. Gilmore looked back over his shoulder. The plane was about a mile away, dropping fast, flaps down, wheels extended beneath the fuselage.

The 707 thudded down on the runway with a puff of gray atomized rubber and screeched toward them. It shuddered, abruptly slowed, and rumbled past them. It turned left onto the taxiway and nosed to a halt as the jeep came up to the forward cargo hatch. They waited until the roaring engines subsided.

The pilot rolled back the tiny side window high up on the hull of the plane. "What the Christ is the diversion for? I got two hundred enlisted and a dozen senior officers on this crate!"

"Highest national priority," Gilmore shouted.

"Oh, my ass," the pilot grumbled. He rolled the window shut and a few moments later, the hatch on the forward belly, just behind the nose wheel well, opened inward like a Martian spaceship. An inky hole appeared in the smooth skin of the plane. A man's head appeared, upside down, then the full figure somersaulted to the ground.

"Th' pilot said sompin' 'bout uh unscheduled stop heeuh," the man drawled. "But the heavy brass ah got topside theah ain' gonna lack this, nossir." He was small and wiry, with frizzy red hair and long, non-reg sideburns. His flight jacket announced he was the crew chief.

Gilmore got out of the jeep, followed by the air policeman with the shotgun, Holt, then the sergeant. The crew chief's eyes riveted on the scattergun. "Okay, fellahs," he said amiably. "You all go anywhere aboard this heah craft you got business in. Ah doan' wanna see that thing pokin' my way."

He jumped back into the plane and they followed, pulling themselves through the hatch and into the murky cargo hold. Red lights glowed dimly and an occasional neon tube flickered in the huge bay. The cargo, crated, palletized, some of it in huge, semicircular containers like pie wedges, filled the hold, giving it a close, claustrophobic feeling.

"Up heah," the crew chief said and climbed quickly up an aluminum ladder into the passenger compartment. Gilmore and the group followed, emerging in the narrow passageway behind the crew cabin.

"I'm the steward," said a chubby warrant officer. "What you want?"

"An enlisted man on this flight. His name is Ronald E. Whistler. He'll appear on your manifest traveling under thirty-day leave orders," Gilmore said. "He's wanted by a federal criminal task force. We're going to take him off this aircraft."

The warrant officer hesitated, looking for direction to the small, first-class seating area where a dozen senior officers, including two generals, sat waiting with a combination of interest and irritation. But they seemed to be waiting for someone else to decide how to deal with the raiders.

A large, barrel-chested Air Force major general suddenly appeared from the small walk-in galley that divided officers' country from the enlisted compartment. He had a grizzled, bulldog face, deeply seamed, lit by a veined red nose. His hair was shaved close on his broad head. He chomped a cigar stump. His service blouse was unbuttoned at the top and the tie was pulled down, exposing a luxuriant mat of curling black chest hair, thick and shiny as a muskrat's pelt.

The general had a .38 caliber Colt automatic pistol strapped beneath his arm.

"What in hail's goin' on roun' heah?" he growled.

The air policeman braced against the bulkhead. "Sir!" he said, clicking his heels and saluting awkwardly, the shotgun jammed muzzle down against his left leg.

Gilmore shoved forward and flashed the federal buzzer in his wallet. Heavy storm warnings were flying in the general's darkening face. Gilmore put a clamp on it: "General, you got a planeload of heroin and a bucket of trouble."

The general's thick eyebrows crawled up his forehead like two furry caterpillars. "The hail you say!" he snorted. "Not on this heah plane they ain' no potheads!" He ducked his head back into the small galley. "Say, Colonel," he drawled to someone inside. "put down that booka war plans an' come watch the fun." He laughed a wheezing rumble like the springs of an old truck. He unlimbered the automatic. "Carry on, son."

Gilmore glanced inside the galley as they went past. A dyed blond with stunning legs was pouring a split of Four-Star Hennessy into a plastic glass filled with ice. "Evening, Colonel," Gilmore said. She turned and flashed a brilliant smile. "Hello, there."

They went into the enlisted compartment. Row upon row of bored enlisted faces stared as the raiding party moved along the aisle. The steward and the air police sergeant checked each name tag as they went aft.

Gilmore saw him before they arrived. Whistler was sitting in the aisle seat of the last row, smoking a cigarette. His face—angular, dark, with a small moustache over heavy lips—was utterly relaxed. A poker face.

"Up! Up!" Gilmore commanded, gesturing with his pistol. "Out of the seat!" Whistler got up slowly, putting his hands in the air. He stepped out into the aisle and they backed into a small open area between the tiny bathrooms at the extreme rear of the compartment.

"What's this?" Whistler said, his voice soft and easy. "What's this?"

"Orders and ID," Gilmore demanded. "Don't reach for them, just tell me." He could feel the eyes of every enlisted man in the compartment on them.

"Orders are beneath the seat in my attaché case," Whistler said. "ID's in my jacket."

Gilmore pulled the wallet out and studied the ID cards. They appeared genuine.

"Lessee," said the general, craning over Gilmore's shoulder. He took the wallet over to a porthole and studied it, growling and chomping his teeth.

Gilmore went through the attaché case. There was a copy of *Playboy,* a small bottle of Chanel No. 5, and a packet of mimeographed travel orders for Whistler, authorizing him to go on thirty days leave beginning two days before and directing him to return to his post at the First Logistical Command, MACV Saigon, not later than twenty-eight days hence. It was signed by an illegible hand "For the Command."

The air police sergeant frisked Whistler for weapons while Holt negotiated with the steward to get Whistler's personal baggage located in the cargo deck and unloaded. They could hear the after cargo hatch being opened. They checked around Whistler's seat, then the party went forward. They went back down the ladder to the cargo hold and then dropped out of the hatch to the runway.

"All right," said Gilmore. "Now I need Private First Class Andrew M. Samuels, serial number five two seven, six five, oh seven four three."

"He's back theah," said the crew chief. He jerked his thumb toward the darkened cargo hold. "He's in a coffin."

"Let's get him up here," said Gilmore. "I think he's going to tell us something."

"No way," said the man. "That's a United States war dead back theah. His destination is Dover and that's wheah he's going."

"CHIEF!" The general glared down through the hatch. "You wanna keep them stripes, boy, you'll do what that man says. Now let's git them offa this aircraft the quickest way, which is to quit hagglin' and do like he wants. Hear?"

"Suh!" barked the crew chief. He turned to Gilmore. "Okay, an order's an order," he said, his face impassive. Then he added in a low voice. "The old fart wants ta screw that piece a tail he's got up there in the senior officer's galley. He wants ta get home ta Dover."

The crew chief led the way back through the long fuselage to the rear cargo door. A pimply-faced enlisted man was climbing through it.

"We found that guy's junk," the kid said. "Christ, all that for one stinking suitcase." He was sweating and disgusted. They looked out. A small yellow-and-blue-striped truck with a hydraulic platform was just pulling away.

"Hey, you dumb trucker!" the crew chief yelled, waving his hands, beckoning the vehicle back. He didn't seem to notice what he was doing. He was absorbed in the story of the general and the general's girl. "You can imagine how pissed he was when we was diverted. The damn flight was late getting under way anyhow. You know why?" He didn't wait for Gilmore's reply. "This Andrew M. Samuels character. That's why ah know who he is. He was late getting aboard. At first they couldn't find him.

"Samuels was supposed to go out on this flight, but the

morgue at Travis said they couldn't locate him. He'd been stashed there for a while, or sompin, and the flight was late. The general . . . Jesus, he was steaming. This is his last flight, this one heah. He's going to his retirement, and that girl, well that's his retirement hobby."

The crew chief was searching around in the mounds of cargo, scrambling over and around it like a large pack rat inspecting its nest. Finally, he gave a grunt of recognition, called to Pimple Face for help, and together the two men pulled a long metal container from the webbed collection of crates and boxes near the rear door.

"Ah knew Samuels was in here," he said. "We had to pack him way back in that hole back theah."

They heaved the container down the narrow aisle and pushed it out onto the platform. "Okay, he's all yours," the crew chief said. "All's you gotta do is sign on this heah manifest so they know who receipted for the item."

Gilmore scrawled his name. The platform truck moved slowly away with Gilmore riding next to the container. He motioned the driver toward the jeep, which was pulling into a small hangar tucked in behind a cavernous maintenance shed.

Flight nine-eight-two's engines screamed to life and the plane trundled back down the taxiway, turned, and roared off into the sky. The truck pulled into the mouth of the hangar and then rolled slowly toward the rear, where a small group of men stood waiting. The truck stopped and the driver lowered the platform. It came down to waist height. Gilmore jumped off and beckoned the air police-man. Together, they slid the container from the platform and laid it gently on the ground. The truck driver yawned, started up, and backed away, his exhaust leaving a bluish gray carpet of fumes on the hangar floor.

"Here's his stuff," Holt said, gesturing with his foot at the ransacked contents of Whistler's suitcase. "There's nothing at all of interest."

Whistler smirked, then his face became impassive, unreadable. His uniform had been taken off; he was standing in skivvies and socks. "I want a lawyer," he said. His voice was relaxed, unruffled.

A jeep drove up and the base commander got out and walked over. "What's the verdict?" he asked. He eyed the shotgun and looked uneasy. He looked as though he had lost his nerve. Gilmore pulled a face at Holt.

"We're about to finish this, Colonel," he said.

The colonel's gaze took in the rifled suitcase and came to rest on the transfer case. "What's that?"

"A transfer case," Gilmore said. "With a body in it. It came with this man here." He gestured toward Whistler.

"Is he the escort?" the colonel asked.

"No," Gilmore said. "This body had no escort, as such."

"Then why's it here?"

"I think Whistler can tell us that," Gilmore replied.

Whistler shook his head. "I want a lawyer," he said.

Gilmore handed a sheaf of orders to the colonel.

"I don't like any of this," the colonel said. "For Christ's sake, this is Andrews, not some podunk airport. This is the President's airport."

"This is the President's task force from the President's city," Gilmore said. "Now let's just go ahead and get this over with, then we'll be on our way." He gestured to the air police sergeant and the man began loosening the series of snap releases along the lid of the transfer case. When he lifted the last one, the lid sprang up. The pungent smell of formaldehyde wafted out. The sergeant pulled the lid up, exposing the insides. They looked in. A long, dark green bag made of heavy gauge plastic, carefully sealed with tape, lay inside.

"Jesus," the colonel said in a low breath.

"What'll we do?" the sergeant asked.

"Lift it out and look in the case."

"Okay. I need some help," the sergeant said, looking at the other air policeman. "You take the feet."

The circle of people around the transfer case drew closer in the gathering gloom of the hangar. Whistler turned and stared.

The air policemen gingerly lifted the shrouded form from the transfer case and laid it on the dirty concrete floor. The colonel sighed as the shape touched the ground.

They moved forward and stared into the transfer case. It was smooth metal. "There's no place in there for heroin," Holt muttered.

"Let's look over there," Gilmore said, pointing at the body.

"Hey," said the colonel. "You can't do that. These orders say the remains aren't suitable for viewing."

"We're not paying our last respects, Colonel. We're looking for heroin."

Gilmore knelt, pulled out a small pocket knife, and sliced the green plastic bag open. The odor of disinfectant billowed out, together with another, heavier smell.

"Jesus Christ," the colonel said.

Gilmore cut lengthwise down the plastic, pulled the edges apart, then closed it.

The body had been autopsied. But the stitches down the naked torso were cut. The incision gaped open at them.

"Someone musta . . . gone in there," the sergeant said, his voice a whisper. "That's where they . . . Put the heroin . . ."

"I don't believe it," the colonel said, his eyes bugging, jaw sagged.

"There's nothing in there now," the sergeant said.

"It's just a dead GI," Gilmore finished. The energy was draining from him. "But no heroin. Somebody got it before we did. Somebody went inside that poor sonofabitch and took it out."

As he said the words, he knew instinctively what had happened—some link in the long chain connecting Haverman with Nick Westley had frayed over the weeks and broken, just as he was at last deciphering the setup. The crew chief had talked of a delay at Travis, because the morgue had misplaced the body. Gilmore knew that wasn't so. Someone had just cracked the pipeline. Or else there had been leaks from the beginning, as each subgroup in the chain through which the bodies passed raked off its percentage of the take.

Perhaps Whistler himself had connived to capture this shipment for himself, or been pressured in ways unknown even to Nick or Haverman to betray their foolproof system. Gilmore knew he would have to work on Whistler to break this information out of him. But for now, for the immediate future, the raid had turned into a disaster. Gilmore turned back to the investigation team.

"We've just bagged the biggest, lyingest, crookedest sonofabitch bogus enlisted man it's ever been my pleasure to throw into jail for about fifteen years," he said. "You're some piece of work, complete with fake orders and fake ID and traveling with one of this nation's honored war dead, desecrated by you and the rest of the vermin you run with."

Whistler set his mouth and stared at Gilmore. "Don't pull that with me," Gilmore said. "It won't wash. When the judge finds out that you, Mr. Whistler, have been cutting into the bodies of our honored war dead to get out heroin, why, he's going to shut your ass up for maybe twenty years. Nick knows that. That's why he made you his courier, Ron. But we can talk about that later. You've got all the time in the world."

"I wanta call my lawyer," Whistler said in a low voice. "You got no right to threaten me like that."

"Just get dressed, pal," Holt said.

"We can put him in Prince George's County jail or the central slam in D.C.," Gilmore said. "Which'll it be?"

Holt rubbed his jaw as if thinking long and hard about it, part of a routine gambit they had worked out. "P.G. is probably safer, but probably won't take him. They don't like to handle federal cases. We'll take Ronnie here to D.C. and see if he can handle himself when they start gang-raping him."

Gilmore nodded. He felt played out, stale. They were trying charades on a man whose only offenses were minor —traveling under false orders and impersonating an enlisted man. They had missed the quarry. Laboratory analysis probably would show some traces of heroin in the mortal remains of Samuels, Andrew M., but that was always open to plausible counterattack in court. The worst had happened: the raid had failed. Nick soon would figure out what had happened, would hear that Whistler had been seized, and somehow, he would get word back to his own suppliers. The Typhoon connection would melt away as though it had never existed. It was the kind of defeat that only Noel Walker would be able to savor fully, Gilmore thought.

He had a beautiful circumstantial case and little more. Even the tie between Big Nick and the gunmen at the Guardhouse was circumstantial, nothing more, really, than a chain of associations that pointed at Nick.

Meanwhile, a few more people were now exposed to the government's case, a few more mouths would chatter the sensational news of a plot to smuggle heroin in the bodies of returning dead servicemen. Gilmore realized for the first time the size of the failure. He turned numbly to the men in the hanger and, as forcefully as he could, warned them against speaking to anyone of what they had seen.

And then, as Whistler was being loaded into an air police paddy wagon for the trip to the D.C. jail, Gilmore thought of Siddler. Where would he be? He would be waiting for word on the Whistler raid. He would need to know the results.

13

Siddler lay in the mud of the treeline watching the little figures trotting in his direction. He thought of Haverman and Tran, cursing that the trail was broken and that they would get away, cursing at the little figures running in his direction, unwitting aides to international smugglers. Yet it all seemed so natural—there he was, lying in the middle of thousands of square miles of rice paddy, the enemy running toward him; they might or might not capture him. As usual, things had gotten out of hand by small degrees. By the time you realized the truly awful nature of your predicament, it was too late; you were in it. How many brightly scrubbed American youths would wake up in the middle of rice paddies like Siddler and think, "Holy fuck, how did I get here? What am I doing *here?*" And know that the answer was, simply, that they had taken, one by one, life's necessary little steps? So Siddler cursed and mused for all of thirty seconds. Then fear came.

Fear is a wonderful stimulant. Where greed and lust and love of goodness and all the lesser passions leave off, fear takes over. It can motivate bad students to get better marks than good students, it can fill small-minded men with sudden sunbursts of imagination, turn meanness into kindness and arrogance into quick empathy. And it can make fat little men run very fast.

Siddler had no choice. He called on all the reserves of his big belly and his big breakfast, sprinted to the other side of the treeline from the burning chopper and the men dressed in black who were coming his way, and then he began running up the treeline, his heavy shoes slapping the muddy earth as he dodged trees and bushes.

His mind worked coolly within narrow confines. There wasn't anywhere to go except along the treeline or out into the paddies, and to go out into the paddies would be like making yourself the target at a shooting range. So Siddler kept along the trees, looking for a place to hide.

The .45 was still in his hand, but it seemed light and almost useless out in the wide spaces. Such a deadly weapon in a room, in an alley, or a street, but here it seemed like a plinking toy in the world of real warfare. He counted up his shots, a mental act that was automatic, that required none of his real attention: seven in the gun and two more clips of seven each in his pocket. Twenty-one shots. It had seemed like plenty of ammo to lug around back in Saigon. Here it didn't seem like much at all. The men chasing him could fire that many in quick automatic bursts from their banana clips, could fire a hundred times that many in a few seconds. At ten times the range. At ten times the accuracy.

Siddler stumbled in thick underbrush, struggled, whirled, fell. His face was buried in moist warm sweetness. Grass. He rolled, gathering himself for the spring, but instinct fought off the impulse to run again. The instinct was honed by fifty-two years of familiarity with the Asian jungle and earth. The grass was high and thick—a man could lie there and not be seen by searchers unless they stepped on him. The Viet Cong used the high elephant grass of the central highlands to advantage in laying ambushes. Siddler saw he was in a large patch of the grass, and well hidden. They would have to search.

With his head and heart bursting from the effort of the sprint, Siddler eased his body down several inches into the grass, snuggled his back into the damp earth carefully, slowly gathering the grass around him and over him, trying to move it only fractions of an inch at a time. Then he lay there, looking up into the cool, dark interior of the treeline, looking over the top edge of the grass almost closed above him.

Slowly, imperceptibly, he forced the tension out of his limbs, breathing deeply and steadily but absolutely silently, his lips open in a wide O-shape. He squinted, looking for any motion, and strained to hear the slightest break in the low buzzing pattern of insects going about their daily business deep in the grass.

He heard nothing. He could not hear the men he knew were looking for him. Practiced in the ways of silence, they could be nearby—as close as a few feet away. They would not suspect a white American to be as practiced in the ways of silence as they, and that would be to Siddler's advantage. Siddler gently moved the safety catch back with his thumb and held his .45 pointed upward at a forty-five-degree angle, his elbow resting on his stomach. He kept his knees slightly bent and his muscles carefully gathered, ready for instant action, a quick spring in any direction. He directed his gaze upward, slightly forward: peripheral vision thus gave him command of a three-hundred-sixty-degree horizon.

Five minutes passed. Ten minutes. Siddler's tension was gone, and his breathing returned to almost normal. But he remained at the ready. Suddenly he heard an excited jabber of Vietnamese voices. They seemed to come from more than twenty yards away, and not from the direction that he had come. A bolt of anxiety went through him as he thought of the tracks he might have left. They would have difficulty finding his tracks, but there was a chance they would.

He listened, not able to understand what they were saying. They were coming generally in his direction. Closer and closer. He felt himself stiffening and fought it. Then the voices seemed to go past toward the other side of the treeline. He judged there were at least a dozen men, many of them talking at once.

After the voices faded, Siddler sat up slowly and stuck his head high enough to see six men, spaced at double-arm's-length, methodically plowing through the grass thirty yards away and getting farther away all the time. He lay down and was still.

His chances were bad if they came back his way. But even if they didn't find him, and nightfall came, what would he do? He could move then without being seen, but which direction would he go? He was likely to be killed by enemy or friendly—the problem was that they couldn't tell the difference at night any more than he could tell the difference during the day.

Siddler was pondering his situation when a faint flutter on his eardrum changed the equations. A helicopter. It wouldn't be Tran and Haverman coming back. It would be some kind of a rescue mission for the downed chopper. Siddler could hear the faint throbbing in the distance getting louder all the time—perhaps more than one chopper.

At almost the instant that he took the decisive move to get up, he saw a man's face and shoulders appear above him—a thin, serious, brown Vietnamese face, looking straight ahead, stationary, not seeing Siddler despite the slight movement. The man was dressed in black, silky cloth.

He had a green sweatband wrapped around his head and bandoliers of ammunition around his chest and shoulders. The man gripped with both hands his AK-47 Chicom assault rifle—a rugged, sturdy, and dependable weapon with a comfortable wood stock that threw out slugs on full automatic with a series of heavy, thumping reports, entirely different from the sharp, popping crackle of the M-16.

Siddler could see the metallic glint of the long, banana-shaped clip through the grass. Stiffening his right arm, he aimed for a spot just above the high curve of the banana clip against the black silk.

Siddler fired once, twice, three times—calibrating the upward movement of his own body as he surged from the grass firing with the downward movement of the enemy soldier's body to the earth, calibrating the man's twitches and jumping motions. It was all automatic. Siddler heard the explosive reports of his gun, felt his arm jumping, saw the flashes, and smelled the smoke—but all else was a wild world of pure action, and he never really saw what happened to the enemy soldier. His last mental picture of the man was of someone standing overhead with a kind of serenity in his face. Then Siddler was running fast, and that was all he knew. He didn't need to think about the three bullets he had fired. That knowledge was part of him now, as much a part of him as the knowledge that there were four bullets left in the clip in the gun and eighteen bullets left in all. Eighteen bullets that, clip changes and all, he could expend in twenty seconds. There wasn't much comfort in it.

He ran through the treeline, burst out into the paddy, and began splashing across it. To his left he could see the lonely charred hulk of the downed helicopter with a dozen

Vietnamese soldiers and peasants standing around it. As he ran, stumbling in the boot-high water toward a dry dike, flailing as in a nightmare where your legs churn but you don't seem to get anywhere, he could see those soldiers and peasants begin to scatter and run, too—but not in his direction. In all directions.

Then Siddler saw the pattern of helicopters and realized why he had run this way. They were coming in low over the fields to his right, sizzling along the ground as he had flown with Watson this morning. He could see a Huey Slick leading the way and the Cobras higher and behind, and then suddenly they were over him and past him and he heard the rattle of their guns that seemed to fill the whole hot noon with thunder and he saw the scattering soldiers and peasants falling.

Siddler stopped and watched the Cobras race along strafing the treeline he had just left.

The Slick rumbled in above him, its M-60s blazing, its wind blowing his upturned face. He saw American faces peering down at him as it came down.

"All riiight, you motherfuckers," he shouted. "Come on down and get me."

They came down, he jumped in, and they swept off. Siddler almost slipped and fell out in the scramble, but strong arms held him steady.

"This is your lucky day, chum," grinned a Special Forces captain.

"Gotta radio to Saigon and then get there," panted Siddler. "U.S. Customs, narcotics bureau. I'm in hot pursuit."

Siddler wiped sweat, mud, and blood from his face with his sleeve.

"Well hot damn," said the captain, fixing Siddler with a

steady gaze. "Them ginks shot a hole right through our radio just a while back. I don't give a fuck if you're Jesus Christ Super Narc, you're gona have to stick around while we finish off this here operation."

After the Special Forces chopper let Siddler off at the Land of the Free helipad, he stumbled through the terminal area looking for a secluded place to make some phone calls. He was enraged and exhausted. His grubbiness felt like a living thing clinging to his body. He wanted a cold shower and a triple G-and-T. But the necessity to keep going pressed in on him, now that Haverman would think he was dead. He had to locate the man and cut him down and he had lost almost twelve hours in the boonies with the Green Berets.

He called Shed Sixteen and cautiously asked for the commanding officer. Haverman was not there, said a polite voice. No, the colonel had not been seen there all day. Perhaps the gentleman would call back tomorrow? Leave any message? Siddler ground his molars and hung up.

Why couldn't he just call Army intelligence or the MPs and turn the mess over to them? How would Gilmore ever know that he, Siddler, had blown the whistle? He pondered it hopefully for a few moments. But somehow, Gilmore *would* find out. And then what? He was too weary to think any further. Do what the man asks.

Siddler called the office of Pei, the Chinese banker, but the old man was either out or wouldn't come to the phone. Siddler sat in the booth and fumed and tried to make some guesses. Haverman could be anywhere in the city, or perhaps he had gone somewhere else? How far could a man like that go before the system itself stopped him?

Siddler dialed again, this time to Trager.

"Ah . . . hello, Ralph."

"Where's Haverman?" Siddler roared.

"He's . . . Ralph, I can't talk over this connection. You'll have to . . ."

"Jesus!" screamed Siddler. "Trager, you stay right there." He was bone-weary from the excitement and frustration of the day. And now Trager was the last goddam remorseless straw. Struggling to his feet, he made for the office.

Dried mud dripped from Siddler's clothing as he plopped down in his chair. He jerked the .45 out of his shirt and slammed it down on the desk top. "All right, Trager, I'm here. Where the fuck is Haverman?"

"Ralph, my God, what happened to you?" Trager's voice was subdued, touched with fright.

"Got in a little tussle down there with our friends Tran and Haverman. They shot down our chopper. I was the only survivor. Rescued by the Green goddam Berets. Haverman thinks I'm dead." He grinned despite himself. "He's wrong, pal."

"Jesus," whispered Trager.

"I gotta get right on the trail, Trager. We got 'em by the nuts. Saw 'em getting a big shipment, fifty kilos at least, and then we got 'em for murder."

"Mur . . . der?"

"That's right, pal. Good old murder one."

"You better look at this, Ralph." Trager handed Siddler a newspaper clipping from *Stars and Stripes*.

"What I gotta read this for?" Siddler complained, noticing that Trager looked unusually pale and drawn, even for him. "You see a ghost or something?"

"Just read."

"Got some booze somewhere?"

"I'll get some."

Trager came over to Siddler's desk and rummaged in the bottom drawers while Siddler murmured and read:

SENATOR'S SON SLAIN IN CAMBODIA COMBAT

Stars & Stripes Vietnam Bureau

First Lieutenant Richard Herbert Paulson, son of U.S. Sen. Herbert Roy Paulson of New England, died in a jungle ambush sprung by NVA regulars yesterday as he led his platoon on a search of suspected enemy supply areas in the Fish Hook region of Cambodia.

The platoon, part of the 202nd Infantry Division, was reportedly pinned down and in danger of being wiped out at sunset yesterday. Air drops of supplies were reported in progress and heavy air cover had been established on the perimeter of the platoon.

In Washington, a spokesman for Sen. Paulson said the parents are in seclusion. In recent months, Sen. Paulson has moved from unquestioning support of the President's war policy to one openly critical of the President for being "too soft." The senator in the past two months has advocated "open warfare on the North," saturation bombing of key military and industrial targets, and the use of U.S. troops in support of "any" invasion by ARVN troops across the demilitarized zone or by sea into North Vietnam.

Paulson has climbed to national prominence in recent years and is being mentioned as a presidential candidate in some political circles. Some political analysts have speculated that

if the war continues for many more months,
the Paulson hard line will find wider and
wider appeal to a nation frustrated at finding
a solution to the long conflict.

Sen. Paulson has asked that his son be buried in the family plot in Arlington National
Cemetery. Sen. Paulson's father, the late Adm.
Roy Greenaugh Paulson, served in both world
wars.

Col. Francis Easton, commander of the operation on which Lt. Paulson was killed, said
a fresh assault will be made today to save survivors and recover other bodies.

Siddler put the clipping down. "What's it got to do with
Haverman?"

"Ralph, that clipping is a day old. Overnight, something
extraordinary has happened. That senator has arranged for
a special Ceremony of Rededication—a kind of national
ceremony and funeral for his son. We got a message today
at the embassy."

"Kinda gross, isn't it?"

"The military seems to be cooperating. And the embassy
guys apparently have to go along with it, too."

"It figures." Siddler was getting angry. "What's it all
mean?"

"Well, Ralph, ah, Haverman left on a flight for Washington two hours ago. He left with the Paulson body. He's
somehow got orders designating him special honorary escort
for the body and there's a cloud of reporters around the
whole thing. He's been interviewed once already on radio.
I heard it."

Siddler was stunned. "You didn't stop him?" he roared.

"Ralph, for God's sake. I didn't have anything on him.
I couldn't with those reporters . . . I didn't know he'd tried

to kill you, or anything about Haverman actually getting heroin . . ."

"What flight?"

"It's called A-37. A freedom bird, ah, Honolulu, Travis, and special stop at Andrews Air Force Base to drop the Paulson boy."

"Christ, his bags'll be packed to the brim with heroin."

"Ah, now Ralph," Trager said, starting to back slowly away, "it's worse than that, I'm afraid."

"How could it be!" Siddler was rummaging through the junk on his desk. "I gotta get after that sombitch!"

"Well, it is worse, Ralph. Believe me."

"All right! Jesus, it *is*. HOW?"

"I killed Manes."

Siddler threw up his hands in disbelief and rage. The man was an utter fool. "What's that got to do with all this?"

"He, ah, he talked before I shot him. I killed him by accident. It happened like this, Ralph, I was poking around over at the morgue at Tansonnhut, seeing if I could help you. It was late. The place was deserted. Then Manes came in. I decided I'd, well, I'd try to be more like you, Ralph. So I just pulled my gun and ordered him into the storage room behind the embalming area. He went, too."

"I 'spose he did. So what'd he tell you, Trager?"

"Well, that they ship heroin inside coffins."

"No shit," Siddler said. He felt a tingle of excitement running through his fatigue. This scrawny little specimen had actually discovered a fact, all on his own.

"And that they ship it inside the dead bodies, open them right up and put it inside and that's how they get the heroin through Customs, because even though we check the coffins, who would ever think to actually look *inside* the body, you know what I mean?"

Siddler had been ready for almost anything, but not this. Visions of the cream-colored embalming tables came back to him. Instead of replacing the viscera, they would insert heroin into the bodies.

Haverman controlled the morgue. He and Manes and perhaps one or two others would see to inserting heroin. A simple courier service, using the bodies of the servicemen. "Jesus," he found himself saying, "it's too goddam macabre."

Mark the body, or tell someone at the other end, and they could easily unload it in the States and sell the stuff. It was a closed, private, safe conduit for drugs. All it took was ruthless cunning.

"Christ, and now we've got an international incident on our hands!" Siddler exclaimed. The move by Haverman had been a stroke of genius. "Listen to me," said Siddler with finality. "I'm going after that sombitch on the next flight and I'm going to get him. And you're going to wire all this to Gilmore on the scramble wire for his eyes only. You understand?"

"All right, Ralph. I'll get everybody informed right—"

"The hell you will!" Siddler blazed. *"Gilmore*—and *nobody* else—is to know about this. If you put out a general alarm, every newspaper will have it on page one and the politicians will be embarrassed and our asses will be in a tight bind. That's the son of a presidential contender, according to *Stripes*. Jesus! We gotta move in complete secrecy until we figure out how bad the publicity is." He thought for a moment then said, "We're going out to Tansonnhut now and catch the next bird. That sombitch Haverman loaded maybe fifty kilos of heroin down at Chau Sit today. I'll bet you he's got the whole thing with him."

"That's a hundred and ten pounds, right?" Trager said.

"Right. And it's worth what?"

"Ah . . . twenty million dollars, right?"

"Or thirty or forty, depending on the market. Now you tell me how much you think Lieutenant Paulson weighs?"

"Ah . . . well, ah . . ." Suddenly, as he was absorbed in the calculation, the expression on Trager's face changed. "Oh my God, Ralph."

"You got the picture," said Siddler. "Yes, indeed. They'll have to stuff that boy to the eyeballs with it to fit it all in— and now just think of that funeral in Arlington National Cemetery with all kinds of bigwigs and maybe even the President there, because they've worked this thing up with the politicians and the senator is going to be some kind of presidential contender some day . . . and they find out what happened to that boy's body! Jesus, Trager, it's . . ."

"Ghastly," said Trager.

"Yeah. That'll do, pal."

During the drive to the airport, Trager filled Siddler in on the shooting of Manes. He said he had gotten Manes into the storeroom. "Then I began that cold-hot interrogation technique. I've never tried that before and it was an exciting experiment for me. No doubt you have great experience with it."

"No doubt. What the fuck is it?"

"That's when you come on all soft and easy at first and then when they evade, you scream at them and give them some negative stimulus, you know, so they'll soften up and talk."

"Negative stimulus?"

"Right, Ralph. I fired my .38 just over the tip of his ear, just after I screamed at him. I guess it surprised him a little. He talked."

"I'll bet."

"He told me about the heroin. He was frightened of you, I think."

"Trager, you're a real peach."

"He told me only that much."

"And then you killed him. Christ, you coulda asked who he worked for."

"I didn't mean to kill him. He got balky so I was going to give him a little more of the cold-hot treatment. I changed ears, though. I fired over his right ear."

"And?"

"Well . . . I guess he caught on I was about to fire and he ducked to the wrong side."

"Oh."

"God, Ralph, it was a mess."

"Right there in the morgue?"

"Yes."

"What'd you do with him?"

"Well, I waited a while, that was the worst part. And when no one came, I loaded him into a body bag—they got them all over the place there—and we rode down to the river and I searched for a while for a nice quiet place and when I found it I opened the bag and in he went."

"Christ, Trager, they'll find him."

"I know, Ralph," and Trager laughed a peculiar, high-pitched laugh. "But so what?"

There was no answer to Trager's question. Siddler had heard stories—they all heard them—of mysterious slayings, of how American soldiers sometimes died at the hands of their own buddies, cut down in an unexplained burst of M-16 fire, or blown to bits by a grenade as they slept. Inexplicable casualties on the gray edges of the conflict. Who fired the shots? Who fragged the sleeping officer? Who saw so-and-so last? Mysteries. The vast organization that ran the war, that sought precision and order within the chaos of fighting, would be suspicious. But it had neither the time nor the resources to find the answers. It had other things to do. Trager was safe. Manes had become flotsam of war.

Siddler moved quickly through the crowded terminal, checked at an information area, and located a flight for the States that was leaving in thirty minutes from a distant gate. He was falling farther behind Haverman. For a moment, he pondered calling someone ahead, simply arranging for a discreet intercept of the Paulson body in Honolulu or at Travis to see if what he believed was in fact so.

But he was apprehensive. What if the job was bungled? What if someone talked? What if he was wrong? How intense was the publicity surrounding the return of the senator's son? More mysteries. It was overwhelming. Siddler made for the plane, anxiety suddenly flooding him. There was not another freedom bird due to leave for three hours. If he missed this one, all hope of catching up with Haverman was gone.

Siddler rushed onto the tarmac. Despite his haste, the sight of the plane took his breath away. From the tip of its black, rounded nose, back along the little windows in a neat row like perfect teeth, along the curving droop of its wings gently weighted like smooth branches with two melon-like engines each, and finally back to the sweep of its tail, it seemed elegant and powerful. Older, duller men than the young troops who named these planes freedom birds might call them Boeing 707s—what lack of imagination! After a year of scrambling through mud and jungle and brush and getting sick and dirty and scared, or after a boring year sitting at a desk in some rear area—in both cases dreaming incessantly of screwing the girl at home— you flew home on a real airliner, with stewardesses who greeted you just as though you were flying from Chicago to New York, a real person and not a grunt just crawled out of the muck, with a chatty captain who told you the altitude and pointed out the sights.

"Son of a bitch," Siddler murmured, knowing that he had made it. They hadn't started the engines yet and he could see the last of a line of green-clad soldiers going up the ramp. "Will you look at that."

He walked the length of the airplane with its orange markings and "Carib Airways" insignia decorated with a little tuft of green-and-orange palm trees.

He went up the ramp and waited behind the line of soldiers getting on.

"Hello, sir," said a blond stewardess, eyeing his muddiness.

"Hello, honey," said Siddler. "You're going to have to take one of these green boys off because I'm in a big hurry."

He showed her the special orders handwritten at the flight desk. She frowned and said, "Oh no, that's terrible. One of these poor boys is going to be awfully disappointed."

Siddler went into the plane behind her. Generals and some colonels were sitting in the front by the door and they looked disapprovingly at him. A wave of tension passed through the troops as the stewardess and Siddler walked down the aisle, as if the men sensed what was happening and knew that one of their number was to be sacrificed for this muddy fat man.

Siddler waited while the stewardess negotiated with the last PFC to embark. The soldier gave Siddler a long, dirty look as he picked up his gear and left. Siddler smiled and settled into his seat, much to the dismay of the neatly dressed soldier in the next seat.

As they left the ground with a terrific, thundering roar, the troops all cheered.

Siddler pressed the call button and ordered a double martini.

14

Siddler awoke and looked out the window. They were coming in low over sparkling blue ocean and white beach. He could see Diamond Head and a row of high hotels, and below there were sailboats, small and perfect like toys.

It was a stunning scene that dazzled his eyes with color and beauty and an impression of people caught up in a simpler, more innocent way of life. And then it came to him, the cry of a doomed man trapped beneath the wreckage of a helicopter. In his mind's eye he could see Watson's mouth opening and hear the scream coming forth—"SIDDLER!"

The memory broke like a wave over him. I owe that sombitch plenty, Siddler thought. He's got a part of me and I'm going to get it back from Haverman.

He pulled the .45 out, inspected it, and chambered a round with a heavy metallic click. The soldier next to him goggled while Siddler put the gun on safety and stuffed it back under his arm.

Inside the terminal, Siddler scurried past the brightly dressed tourists, souvenir shops, and hundreds of soldiers with clinging wives and sweethearts. Despite his rush, Siddler took in the scene—Hawaii was truly the wartime heaven, the land of good food, soft beds, white beaches, and, finally, overworked personal parts. Couples who had thought they might never see one another again came into this airport for Rest and Recuperation leaves and then overdid everything for five action-packed days and nights. Then, tired and worried and sad and maybe walking a little bit funny, they would part again with tears and kisses. Some of them would never come back.

Siddler hurried to the military information desk and learned that the earlier freedom bird, flight A-37, had been delayed in its departure due to engine trouble and was now due to depart in two hours. Relief flooded through Siddler.

"That's the plane the body is on?" he tried.

The clerk looked at him brightly. "Body? Yes sir. That's the one with Lieutenant Paulson on it. Right down there at Gate 17."

"Great." Siddler moved off slowly, scanning the terminal. Haverman might be anywhere. It made Siddler jumpy and furious to be so exposed. But he went on, seeking the one place he knew Haverman eventually would have to be.

He went past the nose of the plane; the tip almost touched the glass of the corridor. It was like a giant's face peering in. A newspaper rack stood in the corridor. "DEAD HERO TO LIE IN STATE" said a headline on the *Honolulu Advertiser*.

The gate area for the plane was jammed with a crowd that spilled out into the corridor. There were flashes of strobe light, and the long, steady brightness of television floodlights. A murmur of voices and the flat, amplified tones of authority. Siddler pushed through the glass door of the reception area and stood at the rear of the crowd that had gathered for some sort of press conference. He strained on his tiptoes to see the floodlighted area up front, where the Gate 17 check-in desk and lounge entrance were.

A one-star Army general was standing there, orating in the full glare of the lights, his face and forehead gleaming with sweat. He stood ramrod straight in his green uniform with bright medals on his chest, referring to papers clutched in his hand, but mostly looking into the cameras, eyes glint-ing and crinkled against the strong light. The general was speaking about "America's liberties." A medium-sized Amer-ican flag had been placed on a standard beside him and two sergeants, leathery-faced and impassive, stood at braced attention on either side of the brigadier.

Behind and slightly to the right was Haverman. The col-onel's hands were clasped in front of him and his head was slightly bowed as he looked down at the floor in a reverent posture. His jowls flexed in the hot glare—but that was his only movement.

Siddler's pulse pounded in his temple. He ducked a little and peeked at Haverman over the shoulder of a journalist dressed in flared purple levis and a loud luau shirt, who was taking notes fervishly as the general spoke with a voice of command into the media microphones.

". . . And so it is with a heavy heart that I offer up this sacrifice to a greatful nation, a weary nation, a great na-tion," said the brigadier. "And so it is that, as we all hope and pray, this tragic war will finish, our great democracy will once again shine bright in the world and lead toward a new peace among all peoples and all nations."

The general looked up and scowled. "That, gentlemen, concludes the statement of Senator Paulson forwarded to CINCPAC here and authorized this day to be read to you by me. There are copies of this statement available at CINCPAC Headquarters and they are authorized to be re-leased at this time."

"You don't have any copies here?" Half a dozen voices spoke at once. "What did the man die of?" "General, can you tell us the exact extent of his injuries?" "General, did the White House say if the President would attend the fu-neral?" "What's the exact arrival time in Washington?"

"General, who authorized the congressional Medal of Honor for an officer whose whole platoon was wiped out?" "General, can you explain the apparent discrepancy between the enemy rifles captured on the Paulson operation—just three —and the estimated enemy killed of seventy-five? . . . Could you tell us more about that? Who made the estimated body count? How does something like this happen?" "General, what is the cost to the public of this special . . . ah, arrangement, for Senator Paulson's son?" "Can you tell us in military terms how important the Paulson operation was to the war?" "Say, there, Colonel, could you spell your name again?" "What outfit you with, sir?"

Siddler saw Haverman's head coming up. He pressed against the bright luau shirt; he could smell the journalist's nervous sweat, see the man's arm working rapidly as he took notes. He could hear a gruff, low voice spelling, "H-A-V-E-R-M-A-N." As the general began parrying questions, a cameraman shouted over the din, "GENERAL! Could you hold your chin up a little and turn this way more, into the lights! That's it—thanks!" Siddler felt as though he were in a school of sharks.

Jesus, he thought, they don't care about anything, really —all they care about is getting information. Those bastards, they'll get it any way they can.

It was a terrifying prospect to think that these men and women might get the real story here, might somehow dig under all the superfluity and really find out what was in the coffin. It would be a national scandal of immense proportions, and they would use it to embarrass the President and that horse's ass, Senator Paulson.

Siddler knew some individual journalists from Saigon and elsewhere in Asia and he liked them very much. They were detached, nosy, irreverent, and frequently very funny

—they were his kind of people. But in a pack, hounding after the facts, there would be no quarter granted. He feared them here in this setting. But because he felt he understood them and knew them, he would beat them. They would find out nothing from him.

There were two immediate problems he could think of: keeping Haverman from killing him and keeping Haverman from the Paulson body.

Neither problem held much chance of solution. As he thought about it, his spirits started to sink. Not only must Haverman be blocked, but other members of his conspiracy must be found and caught. And evidence would have to be arranged and presented in an air-tight case. You couldn't screw around with the solemn rites attached to the senator's son.

When Siddler put it all together in that fashion, it seemed impossible and he knew why he seldom bothered to put everything together. Thinking was as much a curse to man as a blessing.

He listened as Haverman's heavy voice explained how he had detailed himself to be the special escort for Richard Paulson because of the honor he felt it to be to the "American fighting boys in a tough, dirty fight." Haverman told how his offer had been accepted by the Military Assistance Command and cleared "through all the proper channels, up to and including, ladies and gentlemen, the highest echelons of government." "Are you saying that the White House specifically approved you as the escort of his body?" asked a woman. Siddler could see her profile, the attentive thrust of her face, her eyes glittering as she pried away at this secondary fact, hoping to turn it into some sort of "exclusive."

Siddler's mind returned to Haverman's dilemma: how

would he get at the body? If Siddler could dope that out, then he could block the colonel without ever having to expose himself to Haverman. And Siddler very much wanted not to be discovered by Haverman. He wandered down the corridor, staring absently at the plane bearing the Paulson body. It was just like the plane Siddler had flown in on, right down to the orange-and-green tuft of palm trees painted on the tail. But when he had gone thirty or forty feet, he saw something that brought him to an abrupt halt.

A long, sleek, shiny-black hearse was slipping along the runway toward the tail of the freedom bird. Siddler gazed at it—partly because it seemed so incongruous there amid the trucks shuttling baggage, cargo, and food; and partly because hearses carry dead bodies and Siddler's mind was preoccupied with that subject.

The hearse glided up almost underneath the tail of the giant airplane and stopped. Siddler could see it, though it would be out of view of the crowd in the gate area.

Nothing happened for some moments. Then two large black men got out of the hearse and looked around. They were dressed in formal morning coats—the garb of morticians, foreign service officers, and those about to commit the rites of holy matrimony. They went to the tail of the plane. Siddler watched as they produced some papers and began talking with an airline official standing near the plane's rear cargo doors supervising the comings and goings of the transport trucks. The three men conferred at some length. Then the official stepped back, shrugged, and turned to shout something to some baggage handlers.

Before he could see them respond, Siddler had started running. He steamed down the corridor, nicked an old lady and almost sent her sprawling, and scrambled down a short flight of stairs, cursing himself. You gotta go for the nuts, Siddler's head thundered at him. Cut the crap, lard-ass. Sombitch Haverman is pulling one, right in front of my goddam eyes.

"Hey bud, you can't—"

Siddler, his .45 drawn, charged past an official and scuttled out onto the service ramp where the plane was parked. As he ran along the side of the plane, he could see an automatic pallet coming down from the cargo doors at the rear, bearing several stacks of luggage, some small crates, some Coca Cola cases, and a gleaming metal shape covered by an American flag.

The pallet descended and stopped as Siddler ran up. The morticians were starting to slide the coffin off when a surly voice said, "Now wait a minute there." Siddler jammed the .45 back under his arm, and, wheezing, walked around beneath the plane to the other side. The morticians were eyeing someone uneasily.

The voice belonged to a narrow-hipped, wizened little man in a white uniform and frayed black cap. On the cap was a filthy gold braid and an official tag like a conductor's —"U.S. Customs." He was trailed by two younger men, one the airline official Siddler had stormed past and another man, also dressed in white, who surveyed the scene with disinterest.

"Who do you think you are, boy?" the wizened man asked. "You can't go running all over this airfield like it was a baseball diamond." He gouged at his nose angrily.

Siddler realized the man was talking to him. He glanced toward the front of the plane at the glassed-in gate area, where the press conference was still under way. Whatever happened here could not be allowed to escalate—even a whiff that there was some sort of altercation over the Paulson coffin would bring the reporters running. Then the whole thing might unravel.

Siddler fought down the instinct to unload on the old fool. Far wiser to use him. He pulled out his wallet and showed the identification to the official. He held his thumb over the part that said he was a special narcotics investigator. "These guys," he growled, nodding at the hearse drivers, "ain't authorized to get this coffin. They're trying to get it, and I'd shore appreciate any help you could give me, pal."

The old man's expression changed. He was in the company of an ally. He glared at the two blacks, who were exchanging intense, dangerous looks. "Lemmie see your orders, boys," he said.

"We got authorization," said the driver, handing over the papers. "We wouldn't be here without them."

The old man examined the papers, handed them to Siddler, and then went to look at the aluminum transfer case. Siddler read the authorization, a standard military form permitting civilian morticians to dress up a body at stopover points while it was being shipped. It was signed by Rupert Kaiser Haverman.

"Where'd you get these?" he asked.

"Fella brought 'em in the mail," laughed the driver. "What you think?"

"Well, it must be a mistake," Siddler said. "Somebody fouled it up. This man is headed for Arlington National Cemetery and there ain't nuthin' in this world that's going to prevent that from happening." He glanced at the Customs man, who nodded vigorously. "Nuthin'," Siddler repeated for effect.

"This is the one, that's for sure." Siddler completed the

charade by walking over and looking gravely at the plaque on the metal case:

FIRST LT. RICHARD HERBERT PAULSON 436-99-6754
U.S. MILITARY ASSISTANCE COMMAND—VIETNAM
REMAINS NOT SUITABLE FOR VIEWING

"But the papers authorize cosmetic work en route," the driver said.

"Yeah, I know, I know," said the Customs man irritably, "but it don't say a goddam thing about taking him off this plane."

"But how we gonna work on him if we don't run him over to the morgue for an hour or so?"

"Well, goddamit, that's your problem," the Customs man said. "You ain't takin' this fella off, that's all I know." The two men shrugged and went back to the hearse, where they watched as the hydraulic lift raised the flag-draped case back into the belly of the airplane.

Siddler took down the license number of the hearse and imprinted a usable description of the morticians for his notes later. He thanked the Customs man and trudged back through the heat. The plane seemed a mile long. Siddler felt burdened with the knowledge that Haverman somehow controlled far more than he, Siddler, did. At this point Haverman could attempt almost anything on his own initiative and Siddler would simply have to count on luck.

Screw it all anyway, Siddler thought, I'll let Gilmore do the worrying—he's such a smart-ass, he'll know the answers. Siddler went in search of a bar for refueling.

With nine ounces of gin humming in his head, Siddler left the bar and headed out into the main terminal. Still a half-hour before departure. He went along the upper deck, and soon saw Haverman, seated down on the main floor, apparently being interviewed by some journalist. Haverman appeared to be talking animatedly, gesturing with his hands, pausing for a query, then continuing. It obviously was a complex yarn. He ain't hearing a word, Siddler mused, rocking slightly over the rail above. That sombitch's thinking what he's gonna do with that twenty million.

Haverman abruptly broke off and looked directly up at Siddler, perhaps impelled by that instinct not yet gone from man—the instinct that tells us when we are being watched. Haverman's pale blue eyes seemed to reach out and touch the disheveled fat man.

Siddler, shocked and surprised despite himself, and slowed by the gin, fumbled for poise. The journalist, too, was staring up at him with interest, as if he might be discovering another angle to this story.

At last, Siddler was able to move. He raised his arm and waved a feeble greeting.

Then he turned and started down to introduce himself to Rupert Haverman.

"Attention all Vietnam returnees," the P.A. system blared. "Flight A-37 is now boarding at Gate 17 for departure to Andrews Air Force Base, Maryland, via refueling stop at Travis Air Force Base, California. Final call, please."

Siddler stood in line behind Haverman. They were well to the front. They moved slowly toward the check-in clerk. The clerk shook Haverman's hand and murmured a greeting of respect. Haverman returned it.

"Hey, clerk," said Siddler when his turn came. "I'm a Customs agent and I gotta ride this flight."

"Just a minute," the clerk said in a harassed voice. Haverman studied the clerk, his eyes grave and in some odd way, threatening. Then he looked at Siddler and his face split slowly into a broad grin.

"It's true," he said pleasantly. "I need him with me. He's my body guard."

PART IV

A Ceremony for All the People

15

"Heroin in bodies!"

Noel Walker pushed away from the mahogany desk in his private office in the Executive Office Building. "That's the worst thing I've ever heard!" he exclaimed. "It's horrible." He got up and paced back and forth.

"Who else knows?" he demanded.

"The raiding party at Andrews, the command duty officer at the base, and you. And Whistler," Gilmore said. He had just finished telling Walker of the abortive raid on Whistler the night before.

"I want to know whether anyone in the media has been contacted?"

"No one." It was seven-thirty in the morning. A sliver of sun peeped in at Walker's window.

"All right, we must be absolutely sure this does not get out to anyone—anyone." He hit the desk for emphasis. "It could be devastating just now. Devastating in a number of ways. Where is the suspect . . . in D.C. Jail?"

"Yes."

"I want him out of there. The place is crawling with reporters and lawyers. They're likely to stumble across this man and it might get out."

"It seems remote," Gilmore said.

"Maybe, but I don't want *anyone* to know of these allegations until we either can prove them conclusively or drop the case entirely."

"We can shift him to a county jail or up to Baltimore."

"Not good enough. Those places are the same as here—used to leaking stuff right and left. I want him *secure*. Stick him in the stockade at Fort Meade."

Fort George Meade was the Army cryptographic center, one of the most heavily patrolled military installations in the country. It was a few miles from Washington. "He's not in the military. He's a civilian," Gilmore said. "We might get a civil liberties complaint and those *always* wind up in the newspapers."

"Believe me, when that man goes behind the gate at Meade, he will be out of the picture for days," Walker said. "What you have told me will make every front page of every newspaper in the country once it gets out. But we can't take that risk just now. What do you think the American public is going to think when they hear that we believe heroin has been smuggled into the country inside the bodies of our dead soldiers? What are all the patriotic groups that have loyally supported us in the war going to think when we tell the nation what has happened? The political implications just now are incredible."

Walker went over to a gleaming tiger-maple sideboy and poured himself a cup of coffee. A ruddy glow lit his dusty complexion. "There's been My Lai and Hamburger Hill and, some weeks, almost four hundred American boys dying over there. We're getting out of this war the best we can. The country has settled down a lot from recent years. The fight's gone out of the Left. What do you think our critics and enemies will say when it develops that for the past six or seven months—or maybe years, for all we know —the bodies of our honored war dead have been used as . . . vehicles for heroin smuggling?"

"We're going to have to prosecute Whistler," Gilmore said.

"Of course you are!" Walker said. "The point is, right now is the worst possible time. I'm thinking of the Paulson funeral. It's scheduled in two days at Arlington. The liberal press has seized on this funeral and is giving it intense coverage. And some of the columnists are trying to drive a wedge between the President and the senator over this thing. I don't myself believe the senator is doing it for political purposes. I believe he is a patriotic American, genuinely trying to restore the country's sense of devotion and pride in sacrifice." Walker paused. "That's what I truly believe. I think it'll work, too. I think the idea of a day of tribute to our fallen boys, a recalling of the spirit of past sacrifice, it's a good thing. It'll bind up the wounds."

He sipped some coffee. "But it's delicate. We don't need any story about desecration of the dead right there on the front page with the Paulson funeral. We don't want any diversion from that solemn moment. We don't want anything suggested about the Administration but a spirit of cooperation and understanding of the senator in this difficult time when he is placing the nation's interest above his own sorrow. The slightest hint that we had cooked something up would be used by the press as a flog against us. They'd carve us to pieces. So we're going to put the lid on this, just as we did with Luckett. The public will find out when the public is ready to find out."

The intercom buzzed. "Yes?"

"Call for Mr. Gilmore."

Gilmore got the telephone. "Yes?"

"Dan? Holt. Got an 'Eyes Only' gram here for you. From Trager."

"Read it."

Walker looked drained. The plains dustiness of his face had sunk to gray pallor. He fingered the text of the message from Trager for the umpteenth time. Holt was there now. They had been discussing what to do about Trager's telegram for two hours and gotten nowhere. Walker seemed uncertain and anxious. Gilmore caught a feeling of persecution in him that showed through the strain.

Finally, Walker said, "There will be no talk of this beyond this room. I will not convene the coordinating committee."

"No?" Gilmore asked, recalling Walker's lecture of a few days before about the virtues of working together.

"They must not know about this yet. Axby would leak it to Jack Anderson and other columnists. They must know nothing of this. We must—for supremely important political reasons—keep this absolutely secret for now."

"But you will tell the President?" Gilmore asked, a question in the form of a prod.

"Yes, as soon as the data improves. We have lousy information. We don't know whether there's heroin in the body or not. The information has to be air-tight before I'm taking it all the way to the top."

"What about the attorney general?" Holt asked. "He's got to be informed, doesn't he?

Walker flushed. "No way he's going to get his hands on this," he muttered. "And if he hears about it from anyone, I'm going to know where it came from—you or Gilmore. That's all I'm going to listen to on *that* subject."

"In other words," Gilmore said, "there is to be *no* discussion of this with anyone except the President."

"That's precisely right," Walker said. "And so far, we're premature in carrying this tale to him. We don't know if it's true. We don't know—" he broke off. "Jesus, can you imagine the scandal, the impact on the nation? The Left would come howling after us again. There already are plans for a counter-ceremony of some sort by the peace groups at the Paulson funeral. We're trying to keep the Paulson thing as nonpartisan as possible, as free of politics as the senator will allow us. But if this . . . this supreme desecration becomes *public*—the horror of it is going to demoralize millions of Americans.

"A senator's son, as well as a heavily decorated Vietnam combat hero? What's the colonel's name?"

"Haverman. Rupert Kaiser Haverman."

"Yes," said Walker. "I saw an interview with him on the 'Today' show this morning, just before I came over here. He's been officially designated as the honorary escort and now your man Trager says he finagled that himself."

"Yeah, that's what he cabled," Gilmore agreed.

"Well, as of late yesterday, the President himself conferred his personal good wishes and thanks on Haverman in a message that was released to the media. I don't know if the major dailies picked up on it, but I can guarantee you that out in the heartland, in the small towns across this country where they *care* about these things, this entire scheme by Senator Paulson is being watched with a kind of growing sense of national purpose. It's as if the nation *needed* something like this."

Walker paused again, then continued. "The point is, when the President of the United States confers such honors on a person under these circumstances, there is no way —*no way*—we're going to do anything to make a fool of the Chief Executive or make him the laughingstock of the Left. Whatever we do, will be done covertly, privately, and with as few people as necessary involved."

"That doesn't leave us much to work with," Gilmore said.

"I thought for a time this morning that we could make a

surgical removal, perhaps with the Secret Service or the Treasury Department or the FBI. Those are secure groups with dedicated agents. But *I* am not going to be exposed to derision if Siddler is wrong. And believe me, that kind of mistake is precisely what would come back through the grapevine if we are wrong. If we are right, that's another matter entirely and then we go to the President . . . but we won't know until we look, and there's the danger."

Bureaucratic attack and counterattack was as much a part of Walker's life, Gilmore realized, as it was in the life of any obscure public servant trying to make it to his next promotion. The difference here was infinite, however. One misstep and any chance of regaining his congressional seat or taking a crack at a Senate seat was eliminated, for the White House staff would see to it that a man who embarrassed a president could have no political future. And Gilmore knew that the President's staff—his inner circle of advisers—had distinguished themselves as masters of reprisal within the multiple chains of command and authority that twisted through the vast bureaucracies of the federal executive departments.

"But what do we do if we find the stuff in the body, assuming we can agree on how to go about that?" Holt's face was pulled into lines of consternation.

"Well, if it is true, then things are easy," Walker said. "Then we carry it to the President, or at least as high as it can be taken. We show them what they need to know to make decisions—and chiefly, the decision centers around what to do about Haverman—he is the instigator, it appears."

"That should be fairly easy," Gilmore said. "When we have the evidence, we can do as much or as little with it as necessary to get him behind bars for a good long time. And the initiative is with us."

"What we have to guard against is the possibility that the information is wrong. And so we must proceed in utter secrecy, without telling the bereaved senator or his staff or any other federal agency what we suspect."

The black crepe was visible a long way off, draped over the public doorway of the office of Senator Herbert Roy Paulson in the Old Senate Office Building on Capitol Hill. Gilmore headed for the doorway.

A steady stream of people moved through the marble corridors of the building, but there was greater activity around the main door of the Paulson suite. People, some solemn and withdrawn, others curiosity seekers dropping in for a brief shot of emotion, moved in and out of the office. A camera crew from one of the television networks was stowing its gear in trunks on small handcarts outside the office. Two tourists took pictures of the crew.

An ornate door to Gilmore's right suddenly opened and a tall, distinguished-looking, white-haired man emerged, talking over his shoulder to another man. Gilmore recognized the secretary of defense. He wore an air of carefully restrained solemnity, as if he had just concluded testimony about a particularly awesome piece of military hardware. He bowed slightly to the person behind him and walked away, an elegant shadow against the marble walls.

A smaller man with a square, ruddy face and blond hair shot through the door. The eyes were a startling red-brown, like the eyes of a cocker, radiating charm, warmth, instant sex appeal. They were the eyes of a man much younger than the mid-fifties face.

The face belonged to Herbert Roy Paulson and had adorned the cover of one of the weekly newsmagazines some months earlier under a diagonal yellow banner that had said: "PRESIDENTIAL CONTENDER?"

Gilmore went into the senator's reception office. The room was alive with activity, with three young women answering phones and collating press kits on a long table behind the reception desk. The girl sitting there had a bland, all-weather smile and was wearing a pair of out-of-date glasses of blue plastic.

"I'm from the special federal division," Gilmore said. He flashed his badge at her.

"Oh," she smiled absently. "About 'arrangements'?" She put the word in quotes.

"Yes."

"Then you should see the senator's press secretary, John Rigley. He's in charge of that." She got up. "I'll tell him you're here. It may be a few minutes."

Gilmore took a press kit and sank into a leather sofa. The folder contained a photocopy of the newsmagazine article; an extensive biography of Richard H. Paulson; and a listing of major legislation sponsored by the senator during his twenty-odd years in House and Senate.

The senator was in his third term in the upper house, Gilmore read. He was senior to all but two other senators from New England. He had compiled a record slightly to the right of his party's centrist politics and, until about a year ago, had supported the Administration's Vietnam war effort unquestioningly. But when the President had gone back to Congress the previous fall for yet another special appropriation for war, the senator had changed his views, saying the conflict had gone on "too long, too long, far, far too long," according to *The New York Times*.

The newsmagazine speculated that Paulson's move to the right, couched as it was in language critical of the President, had attracted liberal antiwar elements in the country as well as conservative hardliners. The article noted that Paulson appeared to be achieving some acceptance in the country for his conservative views and that the President now faced the difficult task of heading off rejection from a substantial center-right voting bloc and that he was dealing in a gingerly fashion with the senator.

Paulson's move apparently had touched off some kind of response in the electorate. A Gallup poll placed him fourth in a trial heat with the President.

So, the article concluded:

> Fully a year before the party nominating conventions, the senator from New England, the nation's most populous region and traditional wellspring of antiwar sentiment, has become a new voice in the national debate over the war. Whether he will be able to appeal to the legion of political moderates who in the past have firmly rejected right-wing polemics remains to be seen. But as the war drags on and the losses continue, the country will listen increasingly to new voices that reject the present way of doing things. Paulson's voice is growing more influential each week.

The next page of the press kit was entitled, "But Then, Tragedy . . ." and carried a reproduced item from *The Washington Post* about the reported death of the senator's son and the senator's pronouncement: "He died while on a mission in some unnamed part of an unnamed jungle in an incomprehensible precinct of an unpronounceable province . . . he died in vain, make no mistake of that. I shall do all in my power to see that no other young men will be sacrificed to this folly, this tragic folly."

"Yes?" a voice said. Gilmore looked up. A middle-aged man with a harried expression on his long, sallow face had come into the room.

Gilmore smiled and got up. "Special division," he said, shaking hands and thumbing the federal buzzer. "About arrangements." The man introduced himself as John Rigley.

"Oh, all right. Come into my office." Rigley led him into a paneled office and sat down behind a heavily carved mahogany desk. The walls of the room were decorated with photos of the senator and Rigley with various national political and entertainment celebrities.

"There are some precise arrangements that must be made," Gilmore said, "and only a relatively short amount of time to make them."

"What do you mean?" Rigley asked, fiddling with a pencil and pad.

"Customs must inspect everything that comes into this country, as you know," Gilmore said. "Although the flight on which Lieutenant Paulson is being transported will have stopped in both Honolulu and California, the orders for the senator's son bring him straight to Andrews Air Force Base as the first and only official stop."

"I know that."

"I am authorized to tell you that Customs will make its required inspection as brief and simple as possible."

"Inspection of what?"

Gilmore paused before answering, and asumed a grave expression. The bluff had to work. "The body, Mr. Rigley," he said.

"But the Pentagon told us the body was not suitable for viewing," Rigley said. "I don't see how you can—"

"We have trained inspectors who are expert at quick,

positive identification. I'm sure you understand the reason for this."

"I'm not at all sure I do," Rigley said. His desk phone buzzed. He picked it up and snapped, "What," listened briefly, and then said in a mutter that wasn't low enough, "Jesus, no, they can't talk to her. Edna's unreliable, they ought to know that by now."

Gilmore remembered: Edna. Edna Paulson, the senator's wife.

Rigley slammed down the phone.

"The government has two aims," Gilmore continued. "One, to inspect everything that comes into this country, regardless of what it is, for possible contraband—"

"Surely, you aren't suggesting . . ."

"Of course not. I didn't say that. I was simply telling you what motivates the government in these, ah, sensitive areas. There is a second reason. It is to insure positive identification of the body. I'm sure you've heard of mixups in bodies that occurred in World War II and the Korean—"

"But that's not the case here!" Rigley broke in. "This is a senator's son, not some anonymous dogface!"

"I appreciate your position," Gilmore said smoothly. "You must understand the service's. We can't have two standards for people, one for the senator's sons and the other for everyone else. You can see the danger in that, a dangerous charge of favoritism . . ."

Rigley waved his hand in annoyance. "Senators get special treatment every day of the week. There are only a hundred of them and some of them, not more than a handful, will be presidential contenders some day. They aren't like you and me, Mr. Gilmore. They are enormously powerful men, important to the Republic."

Rigley's buzzer went off again, touching off a short conversation that centered around whose name should be put at the top of a particular press release.

Gilmore couldn't tell whether Rigley had won or lost, but the tangle seemed to take some of the steam out of him. "We've all got our problems," Rigley said. Then he added: "I don't mean to hassle you on this, but these arrangements are complex and delicate and involve the world of senatorial perquisites. I don't mind telling you this ceremony of the senator's goes to the heart of the relationship between the executive, and legislative branches."

"I'm sure," Gilmore murmured.

"The senator is a member of what the press chooses to call 'The Club' and what we all know is far more important than that . . . the inner core of the Senate itself. These men, because of their long service and great knowledge of the legislative process, frequently work together for the good of the country, despite party affiliation or political philosophy. So when the oppor—when the senator thought it in the country's interest to arrange a national ceremony on behalf of all the young men who have sacrificed so much, why the president *pro tem* of the Senate and the Speaker of the House did an extraordinary thing. With Congress in recess for the summer, these men nevertheless initiated and ordered that the senator's son is to lie in state in the Rotunda of the Capitol itself.

"Normally, if Congress were in session, that permission would be granted by concurrent resolution of both houses. They have done so in the past for presidents, former presidents and important military figures, as you know.

"The plane bearing Richard Paulson arrives at seven A.M. tomorrow. The body will be taken to the Rotunda and a small honor guard will keep vigil. The next morning, the body will be taken to Arlington Cemetery for a brief service in the Memorial Amphitheatre and then interment in the family plot."

Gilmore took notes on these arrangements.

"You should know that the Army Memorial Affairs Agency some years ago limited burials at Arlington because of diminishing space there," Rigley said. "As it stands now, it is limited to servicemen on active duty, retired military, Medal of Honor winners, others who may be eligible by reason of military service, high government officials, spouses, minor children, and dependent adult children of those mentioned and of anyone already buried in Arlington."

"I understood from the clips that the senator's father was an admiral."

"That's right. But the restrictions now are quite limiting. A grandson could not be buried in Arlington except under obscure regs. The senator's mother has done this for the nation—she has granted to her grandson, Richard, her space at Arlington."

"Kind of her," murmured Gilmore.

"Rest assured that Richard Paulson is himself a genuine hero, a Medal of Honor designee. Therefore, under the present terms of entry at Arlington, he could be interred on those grounds alone. But his grandmother made that gesture . . . and so that is where it is to be."

"Mr. Rigley, the last thing I want to do is insert narrow bureaucratic concerns into what is clearly a sorrowful and trying time for you and the family. But there are certain minimum requirements . . . this procedure of which I speak would take perhaps half an hour in a private, out-of-the-way place. That is not much, really. I am sure that you can work it into your schedule."

Gilmore considered: what would Rigley do if he was told

the real reason for U.S. Customs interest? Would he counsel his boss to keep it quiet, or would he want to use it in some way against the Administration? Gilmore himself didn't care which way the decision went. He cared not to be on the wrong side. He said nothing.

"The schedule is tight," Rigley said. "But let's look." He studied a stapled sheaf of papers and said: "Okay, the body comes in at six-thirty or seven tomorrow morning. There may be a brief ceremony. The body will be taken immediately to the Capitol via a solemn motorcade (lights on) from Andrews Air Force Base to the Rotunda, where it will lie in state for the remainder of the day and night.

"The next morning, at seven or seven-thirty, depending on several other things, the body will be taken to the amphitheatre at Arlington National Cemetery. That's the area behind the Tomb of the Unknown Soldier. It's like a little Greek amphitheatre. The national ceremony begins at nine." An expression of frustration and some other emotion —perhaps ambition, Gilmore thought—passed over Rigley's face.

"As an officer with a rank below company grade, Lieutenant Paulson is entitled by normal regs to a twenty-one man honor platoon, with a captain as the commander of the platoon; a four-man color team; eight body bearers; an eight-man firing party for the gun salute; a bugler; and a caisson detail—the wagonlike vehicle for transporting the casket. There is no provision for a caparisoned horse; that's the riderless horse.

"The last part of the plan is unsatisfactory. There will not be a full-dress, mounted cortege from the Capitol Rotunda to the ampitheatre. Senator Paulson's political colleagues have taken a somewhat conservative and really unjustified position on this question . . ." He paused and pursed his lips. "I shouldn't say that. Arrangements for such things as corteges through the city are well-established by various regulations of the Military District of Washington, the Army crowd that handles such things. The cortege which some of the senator's supporters have mentioned as being a fitting tribute to the youth of the nation who have suffered so much . . . that cortege is simply not provided for in the usual regulations and rules that govern these things.

"I can tell you privately and solely in confidence that neither the executive nor the legislative branches have indicated any interest in waiving that regulation or seeing that it is updated or modernized in light of the needs of the nation."

"Yes," murmured Gilmore.

"Now the senator is not complaining, mind you. He is really playing no role in this funeral ceremony beyond a few words. This is meant to be a spiritual, not a temporal, time. At the same time—and this is why the senator is not making a fuss about the ban on a full cortege—he realizes there is a potential for embarrassment, you might say, to the Administration from this simple ceremony. This is not the senator's intention, needless to say. It was conceived as a ceremony for all the people. The press, as usual, has seen fit to speculate about some of these things, but this is not the senator's intention, I assure you!"

"I understand," Gilmore said. He felt an unpleasant sensation crawling through him.

Rigley stood up. "Let's go see the senator and agree on the best way to accommodate you. You're part of our problem now."

He led Gilmore across his office and knocked at a connecting door, then went through, Gilmore following. They were in an inner office occupied by a secretary with a stern, scornful face. "Ah," she said, "the senator's in with that fat clown." She got up and disappeared through a door behind her.

"She means Bert Morgan," Rigley said. "He's a political consultant. The old bitch doesn't cotton to him. She's the Boss's private secretary."

"Come in," the woman said suddenly, her face popping out from behind the door. "And don't you ever call me that again," she wagged a finger at Rigley. "Or I'll tell the senator!"

They were in the senator's inner office. Paulson was sitting on a couch, dark tie pulled down, french cuffs opened and rolled up. He was talking to a heavyset, florid-faced man in a cord suit who rested on a leather chair. He looked like a large white frog, crouched on a leather lily pad.

"What is it?" Paulson said abruptly.

"This is a man from the Special Division," Rigley said. "He has—"

"What special division?" Paulson demanded. "Ford? Chrysler?"

"Customs," Gilmore said.

"Customs needs to inspect the body of your son," Rigley said as though he were pronouncing sentence. It was designed for maximum impact and got it.

"What!" Paulson flared. "Look at Ricky? In his coffin? That's barbarous! They told us he couldn't be . . . viewed . . ." His face softened. He glanced at the mantel behind his desk. A silver-framed photo of a young, slightly different version of Herbert Roy Paulson smiled at them. The set of the face was the same as the senator's: the eyes had a similar disarming, cocker-like quality to them, limpid, warm, reassuring.

"It's for identification purposes, Senator," Gilmore said

softly. Who could blame them for making a circus out of the death of their son, for trying to use it as a wedge for power?

"There must be positive identification, Senator. It's a legal necessity."

Paulson was composed again, his face a collection of planes and bulges that looked as craggy as the land from which he came.

"What's the best time?" Paulson asked Rigley.

"In the morning, after the viewing at the Capitol ends and before the body of your son is moved from the Rotunda to the amphitheatre. At seven . . . that's the best time."

Morgan shrugged. "If that's what your press secretary says is the best time, Senator, it's the best time. He's the guy in charge of arrangements. If he gets it wrong, you get a new press secretary."

Rigley said, "When you look at what has to be done between now and the final interment, Customs can be squeezed in . . . at the other end, just before the national ceremony."

It was said condescendingly, as though the king's counselor had just conferred a favor upon a supplicant. Gilmore felt his temper rising, but he kept his silence.

Now, the officials running this thing would know that at seven in the morning, just a day and a half from now, a team of Customs inspectors would be given full, private inspection time with the body of the lieutenant. He must locate a small room near the Rotunda where they could do it . . . remove the heroin and substitute plastic bags filled with— whatever suited their fancy.

Then, with documented photographs of the evidence and the heroin itself, with sworn statements from Holt and Siddler, and with the welter of information piling up on how

Haverman and Big Nick ran the operation, there would be a strong case for the prosecutors once the tumult and shouting died down over the apprehension of a military hero just after a nationally televised funeral.

"Mr. Gilmore," said Paulson, "I'm sure you understand our hesitance. It is not meant as a slap at you in any way. It is just that . . . that this ceremony has quite unexpectedly become a ceremony for all the people. It is a time of personal sorrow for myself and Mrs. Paulson, but we have been buoyed by the outpourings of confidence and simple acts of faith by the American people in our behalf in the trying days we now are living through. I know you will understand when I ask you to share with me these trials . . ."

The senator's intercom buzzed. "The commanding general of the Military District of Washington is on the line, sir."

Paulson went over to his desk. The conversation was brief and one-sided, with Paulson doing most of the listening. Then he said, "Well, if that's the decision, general, obviously I'm not going to argue with it." He put the telephone down softly.

"That's a hell of a thing," he said.

"The burial site?" Morgan asked with a fat blink of the eyes.

"Yes, the burial site," Paulson said. "Thank you," he said to Gilmore.

"Thank you, Senator," Gilmore said, getting up. "And please accept my condolences over the death of your son."

"Yes, certainly," the senator said. "Certainly. Mrs. Paulson and I thank you . . ." Gilmore and Rigley headed for the door.

"Well, dammit, can you fix it?" Paulson's voice drifted toward them as they reached the door. "I told them I wanted to be no more than one hundred yards in any direction from the Kennedy grave sites. You know what the general offered us? About two hundred yards over hill and dale . . . Now, I ask, Morgan, whether it isn't the fine hand of the President in there? . . . Huh . . . There's no deal if that's the way they're going to be about it . . ." The office door closed.

"There's something I guess I didn't tell you," Rigley said lamely. "The interment site . . ." He framed his words carefully. "The senator thought that perhaps as a gesture to the nation, he—with the full concurrence of his mother, of course—would donate his mother's burial site to the nation and in return there might be, well . . . a swap for a smaller plot . . . but of course, the new one would have to be closer to, ah . . . you know . . ." The words trailed off.

16

Choler. The word came to him again as Siddler settled into his seat next to Rupert Haverman. The colonel's face was suffused with ruddiness. His neck seemed to bulge with knotted muscle and his eyes snapped and glared as he strapped himself into the seat next to the port.

" 'Scuse me," a light voice said. Siddler looked up. "I'm with the Associated Press," said the baldish young man. "Would you mind if I sat with Colonel Haverman for an interview?"

"You sit in your seat, sonny. I'll sit in mine," Siddler growled.

Haverman said nothing. The wire service man paused, suspended over them, the eagerness clouding in his face.

"Beat it." Siddler said. The man backed away in confusion, and in a few moments, the plane took off and they were turning, climbing out over the Pacific, heading toward California.

"We got maybe five and a half hours together from here to Travis," Siddler said. "Then we got another five hours or so from Travis to Andrews. We're going to get along fine, Colonel. Just fine. I been chasing your ass all over Vietnam and now I finally caught up. I like that just fine."

"I'm going to kill you," Haverman said. "I credit you with escaping and tracking me. I admit surprise. But I'm going to kill you."

"You ain't going to get the chance," Siddler said. "That newspaper reporter . . . he's my insurance man, Colonel. So long as those guys are around, you won't do a goddam thing and you and I know it."

"Neither will you. You had your chance in Honolulu," Haverman said. "You stopped my colleagues from getting at the body, but you didn't move against me. Why not? Is the government afraid of some bad publicity?"

"Not at all," Siddler lied. "It's just that I didn't have my net. All I had was my gun, Haverman, and they tell us not to shoot sick people."

"Others must know," Haverman said. "If you didn't try to stop me yourself, it's because there's a reception committee at Travis or Andrews."

"No one knows. I worked it out myself. I'm cutting myself in on the deal."

Haverman grinned at that, his lips pulling taut across strong, firm teeth. "I can't buy that. I've risked too much already to fumble it away to you. All I have to do is bring this off, this one last piece of work, and I'll be set for life. You're not going to wreck that. Do you understand?"

"I don't understand a goddam thing about you, Haverman. You're a screwball."

Haverman grinned again. "I think of myself as a businessman. I didn't make people into heroin addicts. I didn't start the war. I didn't look for all this. It came to me, unasked. It needed shape and some help, and there I was in a backwater of the war, unable to get to the battlefield, and so I helped them because I had nothing else to do. Think of it as a fault of the system. They spent almost half my life honing me as a fighting man, and when the best war in three decades came along. I couldn't get to the front.

"Logistics!" He spat the word out. "You don't need

ninety percent of the crap we had to take care of. All you need is some troopers with decent weapons and a good leader of men. Someone like me. But the politicians wanted something else. The politicians and the bureaucrats and the cowards. So I sat in Saigon. And one thing seemed to lead to another and now I'm a very rich man. I never had much money and I have it now. I'm an honored member of society, thanks to Richard Paulson's getting himself killed. I cut my orders—my own orders getting me out of Saigon—and I got out."

"For thirty percent of the proceeds, I'll leave you alone."

"Nothing. I'm going to have to kill you."

"I mailed myself a letter in Saigon—Saigon to Washington. It's marked on the envelope to be held five days from receipt and then opened. It will be opened by the chief of the service. You know what's in the letter, don't you?" Siddler liked the lie. It had a nice feeling of credibility.

"It names me and Manes."

"That's right. Tells all I know. Once they have that letter, there's no place you can hide. They'll get you."

There was a long pause. Finally, Haverman said. "How much you want?"

"I figure you got fifty kilos packed into that kid. That's worth maybe twenty million bucks on the street."

"I'm not a street salesman. I'm a wholesaler."

"Okay. I understand that. I'll settle for a million."

Haverman laughed aloud, throwing his head back, his teeth glinting dully. "What's to prevent me from getting the money to you and *then* killing you?"

"Too much risk. Maybe there's *another* letter somewhere just as dangerous to you."

"Take my word. I'm not greedy," Siddler said. "I just

want some return for my trouble. The way it stands, no one else knows. That seems to be worth money."

"What about Luckett? Aren't they looking for Luckett? That's why you showed up at the morgue, isn't it?"

"They found Luckett."

"I knew it!" Haverman said. "That's why you started nosing around."

"How did it happen?"

"Luckett? He was down at Chau Sit and I gave him a ride back with me. I pumped him. He was suspicious, but too young or something to really suspect me. I . . . well, I delivered him to the morgue. Manes took care of him. Just like that. He disappeared. We were safe. But something happened. There was an influx of bodies that night from the fighting up in the A Shau Valley. By the time the morgue was cleaned up, Manes had got Luckett and a kid named DiMalco mixed up. He didn't know whether Luckett had gone into that transfer case or not. By the time we narrowed it down and discovered what had happened, the DiMalco case had already gone out. I was worried because, Christ, DiMalco, well there wasn't much there. If anyone paid attention, they'd see the difference in weights. I checked and DiMalco was headed toward the Washington area. I passed a message to an associate there, to keep his eyes open, look in the newspapers for anything that might tell us whether we were in trouble. It was the only chance I had. Meanwhile, I started closing out the operation.

"When you showed up, I knew what I had to do—I had to get the last and biggest shipment out." Haverman paused. "Now I've done that. We're safe. You and I. Absolutely safe. I haven't heard from my Washington friends, so I have to assume the police haven't made any connection to them yet.

I think we're in good shape. But I still don't know how Luckett led you to me."

"He didn't. The bank account did."

"Oh, yes, the Chinese. You can't trust the little bastards. Well, my bank account has been closed out in Hong Kong. It's on its way to Switzerland."

Haverman ordered a brandy and soda. Siddler ordered a bloody mary. "The closer we get to Andrews, the better things get for us," Haverman said. "The press is covering this funeral like it was a real story. I've given three television interviews already." He shook his head. "It's very funny."

"What about the money?" Siddler said.

"What money?"

"My money."

"I think a million dollars American is much too high, even during an inflation. You'll have to come down, and I've already assumed you will because I'm considering you a partner. How about two hundred thousand?"

"Half a mil."

"Three hundred"

"Four hundred."

"Three hundred."

"Three hundred"

The drinks came. When they got them mixed, they raised their plastic cups. "Here's to a successful, brief partnership," Haverman said.

"Not too brief, pal. Not too brief."

They had a long lunch and drowsed through an in-flight movie. Siddler shooed off another journalist with the promise that they could interview Haverman when the plane reached Travis.

Haverman pushed the seat back and settled himself for

a nap. He turned on his side, facing the porthole, and curled down. Then he turned and said to Siddler, "You wouldn't be thinking about taking the whole thing, would you?"

"Yeah," Siddler said. "I'm thinking about it all the time. Problem is, I gave my word. I wouldn't lie to you."

"That's what I thought," Haverman said.

"Well, it's all you got to go on, pal."

"I know." Haverman turned away and soon was asleep.

The colonel's back rose and fell gently, the air moving softly through his generous nose. His sleep was stirred by an occasional twitch of his left leg or a subtle working of the jaw muscles. He's chasing rabbits, Siddler thought.

Was he a rabbit to Haverman? Yes. The answer chilled him. There were parts left out of Haverman, making him a simpler machine than his competitors, ruthless and uncaring, quicker and far more deadly.

They ought to have a laboratory full of guys like Haverman so they can study them and tell the rest of us how to recognize and protect ourselves from them, Siddler thought. Then he thought: No, they would have to give each of us a gun and train us how to hit a moving target. That was the only way to protect yourself from the Havermans of the world.

I could take my chunk of steel out from under my arm, cock the sombitch, and blow him away, Siddler thought. I could save myself a lot of worry and the taxpayers a hell of a lot of money. I might even save myself from getting killed. And once we showed our evidence to the prosecutor, there isn't one in the country who would press charges.

So why don't I? He pondered it. Haverman was owed one. Siddler felt a sensation in his right hand . . . he could calibrate precisely where the hand was and where it should move to grasp the handle . . . it would be a single, smooth action, slide the pistol out, cock the hammer, and blow his brains out. A third of a second? A half? He thought about it for a long time. Haverman sighed in his sleep and faced forward. One pale blue eye opened, focused on Siddler, then winked shut. Siddler could not bring himself to do it. But he knew he would try, given the slightest provocation. He knew he wanted to kill Rupert Haverman.

The plane reached Travis Air Force Base at seven P.M. It was greeted by a small group of television cameramen and some reporters. Haverman was surrounded by a half-dozen public relations men from the Pentagon and together they conducted a brief news conference. Siddler was introduced by Haverman as a man from the Saigon embassy. Then they went into a private lounge area off the officers club and had some drinks. Siddler went along and listened as the Pentagon men described the dimensions of the Paulson funeral:

It was an important occasion, the Pentagon had decided, and without blowing the ceremony out of proportion to the other things the country had to concern itself with, the funeral would give the military a chance to join with an increasingly powerful civilian critic of the war and make him see the value of strong defenses. "He'll be eating out of our hand by the time it's over," said one. Therefore, there would be special arrangements for an all-services honor guard at the Rotunda, where the body was scheduled to lie in state; there would be a full platoon of soldiers from the ceremonial Old Guard Division at Fort Myer to lend special meaning to the funeral and the interment at Arlington.

He, Haverman, was central to the Pentagon's handling of this ceremony. His presence with the body lent a touch of high-ranking dignity from a Vietnam veteran. In turn, the

PR men congratulated Haverman for his devotion to duty and for detailing himself as special escort.

"It's because I admired the kid so much," Haverman said, sucking at a brandy and soda. "His defense of his men was remarkable and served as a lesson to us all out there." The Pentagon men looked at each other and eagerly nodded their heads.

"We were going to put a brigadier on this," said the senior PR man, "but Colonel, I don't doubt that you are handling this in a way that is bringing credit to the service and the nation. You are to be commended for your initiative. And frankly," he said, *sotto voce,* "if we left it up to the secretary's office to decide what to do about the Paulson funeral, we'd still be arguing. Your direct action made it much easier for us all. No one can say that in the time of need for this important and bereaved family, the Army didn't do its best."

They shook hands all around and Siddler and Haverman went back to the main terminal, where the two reporters from the plane got their interviews with Haverman. Siddler stayed right there. He wanted Haverman quiet, off-guard, and with his suspicions quiescent for the arrival at Andrews.

Siddler woke with a start. There was a changed pitch in the heavy whine of the engines; they were softer now and the plane was thumping and whirring as it lowered through clouds. Haverman was reading *Playboy.* Past his shoulder, the sun was just coming up. A few tiny pinpricks of light moved down there. They were beginning their descent to Andrews.

A stewardess came by and said to Haverman, "Colonel, the steward wanted to make sure you had time to ready yourself." Haverman went to the rear of the plane. Siddler ordered a bloody mary. It came in pieces: a plastic glass;

four perfect ice cubes, their edges glistening; a miniature can of V8 juice; a slice of lime, looking as though it had come from a candy dispensing machine; and a split of Smirnoff. The sight disturbed him. "Where's the goddam blood?" he growled, fumbling up the cash from one of his pockets.

Arterial roads and sprawl of housetops were wheeling slowly below as the plane approached the airfield.

You get used to Asia, Siddler thought, where a white man, even a fat, ugly white man, has a kind of status and mystique, and a knowledge of his own rarity that makes his passage remarkable. He tried to imagine himself in one of the cars down there, racing to be first to the exit, first to the parking space, first into the elevator, first out at night, first onto the arterial turnoff, first first first. He dumped the split into the glass and gulped the vodka neat, tucking the lime, juice can, and empty glass into the seat pocket in front.

Then he wrote a quick note on a piece of paper torn from his dog-eared hip-pocket pad. He folded it and stuffed it into his right jacket pocket. He had no idea what sort of reception Gilmore had planned for them at Andrews. But he knew from the briefing at Travis that there would be a small, solemn ceremony at Andrews when Senator Paulson and his wife personally greeted the casket bearing their son. From there to the Rotunda. He could not leave Haverman for more than a few moments, if that. He had to stay with him to convince the colonel of his sincerity.

And what of Haverman? What would he do? He had to make some sort of arrangements to retrieve his heroin. Siddler would have to stay with him. Siddler would have to head him off. He knew there was no way the colonel was going to let him share in the loot. He couldn't let Haverman out of his sight. He must pass the message to Gilmore—don't interfere, I've got it on my own, keep back, and I'll deliver the sombitch to you, COD.

Haverman came back. His hair was glistening and he had shaved; the skin looked smooth and taut. He smelled agreeably of shaving lotion. His eyes were hooded and brooding. He slumped into his seat and muttered something.

"What say?" Siddler prodded.

"Nothing."

"Don't like it down there either, do you?" Siddler guessed.

Haverman nodded. "Dwarfs. They're all dwarfs. What do they know about war, about fighting? Nothing. What does the promotion board know about war? Nothing. Old morons sitting in their clubs drinking as much as you drink. Sickening. Utterly sickening."

"Oh, I don't know," Siddler said. "It ain't all like that. Problem is, you don't belong down there. There isn't room for people like you down there."

"You think not?" Haverman seemed pleased with the notion.

"They'd need a cage around you, or a big bell around your neck, like a cat. They ought to bell you, Haverman."

"Now, look, partner, we aren't going to get along very well if you keep saying things like that. I might change my mind. Five days before they open that letter you sent Customs? In five days I can get clear into Brazil or some place like that. Maybe a Central American country where they don't have extradition agreements. I'm safe."

"Okay," Siddler said. "I didn't mean to upset you, Rupe. It's just that sometimes I think out loud."

"You weren't thinking of putting a bell on me, were you?"

"No," Siddler lied. "I was thinking of how different you are from most people I've met. And how much I'm going to like the kiss of that money. And I was thinking how I'm going to make sure you don't skip out on me and how I'm going to be sure everything goes smoothly."

"How do I know you're as good as your word? Any man who would sell out as easily as you is a bad risk."

"You just happen to be the man I've been waiting for all my life," Siddler said. "But don't forget the letter, Colonel. The letter is what you have to worry about."

"It will take time to make a connection and get the stuff sold once it is, ah, retrieved," Haverman said. "How can I be sure of you?"

"I'm going to stick with you, just like a shadow."

"Right through the ceremony? It's going to be incredibly dull."

"Not with the punch line you've got figured out." Siddler chewed his lip. "How you going to do it?"

"That's a trade secret, partner. Even partners have secrets from each other."

"Then I'll just have to stay with you, won't I?"

"Yes," said Rupert Haverman. "You'll just have to wait and see."

17

Gilmore and Holt took up position at the rear of a small crowd of reporters gathered to observe the arrival of the body of Lieutenant Richard Paulson. The incoming freedom bird landed and pulled over to a loading area near the terminal where a podium had been set up and cordoned off.

Senator Paulson and a number of officials arranged themselves around the podium and as the plane's doors started to open and soldiers began emerging in long khaki lines. Rupert Haverman, resplendent in a green dress uniform, came down the stairs at the front of the plane and marched over to greet the senator with a salute and a brisk handshake. He was followed at some distance by a fat man dressed in unadorned tropical khakis. "That's Siddler," Holt said.

"Why's he staying so close to Haverman?" Gilmore muttered. "Afraid the colonel will shoot him?"

"Maybe he likes Haverman."

Just then, the flag-draped coffin, pulled by enlisted men, appeared around the nose of the plane and was wheeled up to the podium area.

"We'll have to go over there to get Siddler," Holt said.

"That's fine," Gilmore said. "What we need is a little exposure. The TV crews will go right for it."

They pushed forward through the crowd. "Marshals. Federal marshals," they said, elbowing their way along, showing their badges and walking out around the roped-off area. They went over to Siddler, who was hanging back with a pained expression on his face.

"You're under arrest," Holt said in a voice that carried across to the official group. Senator Paulson, who was about to begin some brief remarks, looked annoyed. Haverman strode over. Siddler struggled mildly, angry and embarrassed.

"What in hell's going on?" Haverman asked. His face was flushed.

"This man is impersonating a U.S. government official," Holt said. He worked his cuffs and clamped them on Siddler.

"I've got identification," Siddler growled.

"He's been riding with me since Honolulu," Haverman said. "I don't understand what you're saying."

"We're saying he's a parole violator and fraud," Gilmore replied.

"I haven't done anything!" Siddler yanked hard, breaking free of Holt's grasp. He turned to Haverman. "I don't know what all this crap is. I'll clear it up in a few minutes, Colonel."

". . . and I'd like to express special thanks to Colonel Rupert K. Haverman, of the First Logistical Command, for his special escort duty of the body of our son . . ." Paulson said. "Colonel Haverman, who is completing his second full tour of duty in Vietnam, generously offered his own personal leave time to accompany my son, Richard, home from this conflict . . ."

The television cameras were now concentrating on the arrest scene. Holt and Gilmore dragged Siddler from the crowd and hustled him into a sedan parked near the entrance.

"What's his name?" yelled a reporter. "What's he done?"

"He's got a dozen names," Gilmore said as he started the car. "He's impersonating a federal officer."

Holt looked out the side door. "Christ," he said, "Haverman's bug-eyed." Then he grinned. And said, "You remember me, don't you Ralph? My name's Bob Holt . . ." He offered a hand over the seat. Siddler declined.

"Holt? From Customs? Good Christ, you clowns know what you've done?"

"Yeah," said Gilmore. "We got you away from the bad-asses. I'm Gilmore."

"The hell you did!" Siddler roared. "You dumb shits have screwed the whole thing up! Let me out of this goddam boat."

"Looks like we should have gagged him too, Dan," Holt said.

"Now, listen a minute," Siddler said. "I got to stay with that ape. It's the only way he isn't going to break and run. You'll wreck the whole thing."

"If he runs, we follow," Gilmore said. "It works like this: we're under absolute secrecy on this thing. It's dynamite to the pols if it gets out. We've instructed a team of agents that Haverman is to be protected, unobtrusively, for the entire time he's here; Holt and I have made arrangements to watch the body ourselves while it's in the Rotunda and at Arlington."

"Well that's just fine," Siddler said. "Get me out of these bracelets. And then get me back to Haverman. That sombitch is going to pull something and believe me, we don't want to get caught with our pants down."

"Haverman's contact in Washington has been a warrant officer named Cronder, a conduit for messages. Cronder's in the cooler. Cronder passed messages to a black drug dealer named Big Nick Westley. Nick is holed up in a place where his phone is tapped. He's under twenty-four-hour surveillance. What I'm telling you is we've sealed off Haverman from his contact and his connection. He's isolated. He'll find that out when and if he tries to move. Meanwhile, we've got our own plans to get at the heroin."

Gilmore went on to describe the preparations. "We've got permission from the Capitol police for a cordoned-off area just below the Rotunda, an anteroom to a staffer's lounge that's on the floor below. The body goes down there at seven tomorrow morning. We go with it. The honor guard is ushered out. We go to work—"

"Who's we?" Siddler broke in.

"You, me, and Holt."

"Not me, pal. I'm going to be with Haverman. What makes you think he's going to desert the Rotunda? He'll be right with that body, riding shotgun wherever it goes. That's what I've been trying to tell you for the past ten minutes. Haverman and I . . . we're partners. I've got three hundred thou riding on that casket. It's the only way I could keep the sombitch from letting light into my head . . ." Siddler quickly sketched in the deal with Haverman.

Then he said, "You've queered the whole thing. I was going to stick with him 'til he made his move. Then I was going to lower the boom."

"Meaning what?" Holt asked.

"Kill the sombitch."

"He'd get you first."

"He already had his chance down near the Seven Mountains. He boobed it up. Now it's my turn."

"All right," Gilmore said. "He doesn't know what in hell's happened with you. He doesn't know whether you've been arrested for real—"

"Dammit, it's a charade with him," Siddler broke in. "He wasn't going to keep that 'deal.' It was simply a convenience to keep me around."

"Then we'll turn you loose after you've had time to clean up and get some rest. You can tell him they locked you up by mistake and you got out. He's going to be in the Rotunda with the Paulson family. You can go up there. Holt or I will be there too. The body will not be moved until seven A.M. then it comes downstairs and is under our control for a half-hour. After that, he can have it. We lay back and watch what he does." Gilmore paused. "Either way we've got him. If he tries for the heroin, we nab him. If he doesn't, we produce the evidence and arrest him ourselves. After the funeral."

"He ain't going to believe that yarn," Siddler said. "No way."

"Well, it's all you're going to give him to chew on. He isn't going to plug you right there in the Rotunda, he isn't going to do anything like that, so you're safe and you oughtta quit bellyaching."

"I've heard of smoother plans."

"I admit it—it's makeshift and has some thin spots. This isn't like blowing a safe or teaching your kid how to ride a bike. We're trying to narrow the possibilities to what we can handle when there aren't many of us and aren't going to be any more of us."

Siddler understood. It had to do with probability and how to reduce the options left open to the quarry.

"We've got one other thing on our side," Holt said. "The Rotunda is going to be jammed with people in the early hours. Thousands of visitors, dignitaries, the press—and some veterans groups are making bus tours to the Capitol. It's something of a curiosity if you're a tourist. If you're a pol and you're still in town, you want to be seen up there on the Hill, paying your respects to this boy, this son of a powerful senator. If you're an antiwar type, you'll be there, too. You might be causing a commotion, but you'll be there."

"Does the honor guard know the body's going to be moved at seven?"

"No. We'll tell them about fifteen minutes ahead."

"Good," Siddler said. "Haverman's a colonel. If he got wind there was some plan to move the body, he'd countermand it and they'd probably obey him, not you." He waggled his wrists. "How about getting me out of these?"

Holt released him Siddler rubbed his wrists. "I got two questions, then I'll shut up."

"Promise?"

"One: what's to prevent Haverman from getting at the kid right now? He's going to smell a rat in this and move fast. That's what he did when he found out I was onto him in Vietnam."

"The hearse from Andrews in being driven by an Army enlisted man. An agent from Task Force Washington is sitting in the passenger's seat. There's no room for anyone else."

"Okay. Two: how you going to pry Haverman loose from the Rotunda long enough to get the heroin?"

"That's a trade secret, pal," Gilmore said.

"Not again. I heard that once today. Or was it yesterday?" He looked out the window at the jumble of shops, gas stations, and small apartment buildings that stood back from the road. The realization that this was Washington jarred him. He was feeling the time displacement of the long flight.

"What I'd do," Siddler said, "is tell the Army CID or

somebody like that and let them take care of it. Doesn't anybody trust them?"

"We're handling it ourselves because of the political problems."

"Well, they can do a perfectly good job," Siddler grumbled.

"You know that and so do we. But the guy we work for doesn't."

Siddler grunted. They were talking about a kind of life he hoped he would never need to deal with again—a life of bureaucratic intrigue and organizational suspicion, where every move had a counter-move, every action its vulnerable weak point, where each facet of performance was measured in precise ways against the performance of others. He was not impressed; but he hoped they were right.

They drove to Gilmore's apartment and dropped Siddler and parked the car, then went over to the Capitol. As Holt and Gilmore arrived, the small cortege was mounting the steps of the East Front. Gilmore checked his watch—nine o'clock. Exactly twenty-four hours to go to interment.

A throng stood silently in the solid white flanks and ramparts of the building, watching as the coffin went up the Capitol steps, carried by an honor guard of men from each of the military services. The Paulsons followed. Haverman was next, alone. Then came a succession of politicians—all the congressmen from Paulson's home state; senatorial colleagues from around the country; a number of distinguished citizens of the home state; some Administration middle-levelers; numerous legislative lobbyists from various federal departments; and finally, the President's press secretary, a personal emissary.

An Army band played a slow march. Traffic on Constitution and Independence avenues flanking the Capitol slowed to a crawl, then stopped altogether as motorists watched the ceremony.

The coffin disappeared inside to the slow cadence of the band, the drums rattling and thumping in mournful beat. The crowd of spectators had grown to enormous proportions, spilling off the steps of the East Front, across the parking lot, and well into the park beyond. Gilmore scanned the crowd with interest, discovering that it was largely white, middle-class, conservative in dress, quiet in demeanor, with few hairy faces and antiwar freaks. The crowd surged forward, carrying Gilmore and Holt with it. They were funneled through a narrow entrance to the Rotunda.

The Rotunda was cool and breezy, with muffled voices and shuffling steps echoing dully in the upper reaches of the dome. The coffin lay in the exact center of the huge circle, separated from the slowly moving stream of people by a cordon of velvet ropes. Paulson, his wife, and their two daughters stood across from the coffin, facing the passing throng, their heads bent in silent meditation. Farther back stood the dignitaries, and beyond them, groups of Hill employees moved quietly back and forth from the hallways leading off the Rotunda.

Rupert Haverman stood at the foot of the coffin, at parade rest. His legs were spread, chest out, hands clenched behind his back; he was a ruddy, muscular anchor to the scene. Gilmore and Holt sidled out of the moving stream of mourners and leaned against one of the marble columns.

Siddler let himself into Gilmore's apartment. He made a face as he smelled a distinct aroma. "Pothead," he muttered. He went into the kitchen, found the vodka and mary mix, and poured himself an enormous drink, downed it quickly, and mixed another. Then he rummaged in the fridge and proceeded to cook up all he could find—half a package of bacon, four eggs and four English muffins and a pot of coffee. He took it into the living room. A tall, dark-haired girl with the smooth, elastic physique of a gym teacher was waiting for him. She was wearing a pink peignoir and nothing else.

"Hi," he said.

She nodded and introduced herself. "You've got quite a breakfast there," she said.

"Sure," he said, starting in on it.

"You don't look starved."

"I'm not," he replied around a mouthful of English muffin. Neither do you."

She smiled. He liked Anita's smile, but he didn't think her name fit. That name didn't belong with her large, promising body. He gulped the second mary and looked at her some more as she leafed through the morning paper. Gilmore was all right, to keep such a woman around the house. She looked like a good performer. He bet he could teach her some tricks that Chi-Chi had sprung on him at different times. She'd like them, he could tell by looking at her wide mouth and large hands. But it would never happen. This was a tall man's girl, not a short, fat man's girl.

He drained his glass, rolling the mixture around in his mouth to catch the bits of muffin and bacon caught in his teeth and gums. He watched her and thought: what the hell. So he sat back and said, "How about it?"

She shook her head. "Nope."

"Why not?" he asked taunting her.

"You wouldn't give me any of your breakfast," she said. "So why should I?"

Gilmore stood in the nighttime shadows at the edge of the Rotunda. The bier was lit by golden flames from large candelabra. The flickering light illuminated the solemn faces of the young servicemen who stood at attention at the four corners of the cordoned-off area. The vast space was filled with the sound of shuffling feet, murmurs, an occasional cough as a thinning stream of mourners and spectators moved past.

The day had passed uneventfully. Haverman had stayed at the bier with the Paulsons throughout the morning, then had gone with the family to their house in Chevy Chase, where he was to be accommodated during his stay. Reports from the agent tailing him were being received by walkie-talkie, both at the task force's downtown headquarters and by Gilmore at the Capitol. The agent reported that Haverman could be seen from time to time on the terrace of the house, taking his ease and conversing with Senator and Mrs. Paulson and some of the other houseguests. Haverman had gone inside at four P.M. and apparently was resting. There was no tap on the senator's line so it was impossible to tell whether Haverman had placed or received any calls.

Meanwhile, Nick Westley was at his uncle's funeral parlor. A bug had been placed on the front window of the house, but had been ineffective. The tap on the McVey telephone had revealed nothing. Several funerals had been handled by the parlor and at one of them an agent impersonating a mourner had slipped into the house. He had seen no trace of Nick.

Several times a day, the highly polished limousine owned by the McVey home was driven out. It had been tailed each time, but to no avail. There were no clues to Nick's intentions.

Gilmore strolled to the West Front of the Capitol and

looked out. He was fighting an attack of nerves. Everything seemed too peaceful. The city lay before him, with the vast federal buildings along Pennsylvania and Constitution avenues gleaming through the foliage and the lights of traffic moving slowly in streams around the monuments and edifices. Holt would be relieving him at eleven for the long night-watch. Siddler himself would be calling Haverman soon to tell him that he would be going to the Capitol after being released from jail. Siddler was to decline any invitation to meet Haverman anywhere but at the Capitol.

Someone sidled up to Gilmore. "Hello," he whispered. It was John Rigley.

"How's the senator?" Gilmore asked.

"Bearing up."

"And Colonel Haverman? How's he?"

"We had a small dinner at the senator's house. I just came from there. The colonel has gone to bed, I think. Or is about to. That was quite a trip, across country and all. Jet lag."

"Yes."

"Mind if I ask a question?" Rigley said.

"No."

"Why are you here?"

"In the Rotunda?"

"Yes."

"I'm paying my respects to Richard Paulson," Gilmore said. "Just like you."

"And why are you so interested in Colonel Haverman?"

"Idle chitchat."

"Why did you identify yourself to the press at Andrews this morning as a U.S. marshal when you picked that fat man out of the crowd?"

"Never tell them anything they don't need to know."

"Okay, Mr. Gilmore, have it your way," Rigley said. "But

let me tell you something. I happened to look up the so-called Special Division of Customs earlier today. There is no such thing. Then I got suspicious, so I had a friend at *The Post* make a check. They've got quite a bit on Daniel Gilmore and the narcotics task force. You'd be surprised at how much. Not what gives?"

"Mr. Rigley, I want you to know that there is no way I can answer that question—no way at all."

"If you don't, I'm prepared to go to the senator and tell him of my suspicions."

"And what are your suspicions?"

"That the task force has some information which affects Colonel Haverman. If that is so, then you are duty-bound to reveal it to the senator . . . now . . . when there is still time to do something about it."

"You are one hundred percent wrong, Mr. Rigley," Gilmore said. "I can tell you that as truthfully as I know how. If you carry such a wild allegation to Senator Paulson, you will be slandering one of this nation's military officers. You will bring embarrassment to your senator. It would be foolish in the extreme."

Rigley said nothing for some moments. Then: "All right, Mr. Gilmore, have it your way. I was just offering help, that's all."

At midnight, Ralph Siddler sweated his way up the steps of the Capitol, bathed in pale moonlight. Siddler felt eerie. He had never set foot inside the Capitol of this country for which he had labored so long in such out-of-the-way places, for whose paycheck he had risked his life perhaps five dozen times in the course of his service. The dark shapes of the people mounting the banks of steps on the East Front, and the silent shadows of the guards placed here and there about the flanks of the huge building; the floodlit dome soaring so powerfully above them; and inside, the flickering light of the candles and the profiled faces of the mourners—all of it cast within him a sense of awe and inspiration which he suspected was resurrected from his long-ago youth as the son of a minister. He was suspicious of the feeling, but it was there and strong within him: a reverence for the place of power where he now walked.

He felt himself drawn to the bier itself. He joined the sparse line and went slowly past, drinking in the rich scene: the muffled sounds, the pristine military guardsmen, their eyes shining in the candles.

Siddler found Holt along one wall, lounging by himself in the deep shadows of a column.

"Gilmore's gone to bed," Holt said. "He'll be up at six and here by six-thirty. What's with Haverman?"

"The colonel was said to be asleep. I called to tell him I was released. They said they would deliver the message, which they did."

"What'd he say?"

"He didn't say anything. He didn't come to the phone. Strange."

"If you want to know, I'm happy we've kept this thing small . . . only a few know," Holt said.

"Getting doubts that it's in there?"

"Maybe."

"It's there. Don't worry."

The time passed slowly. Siddler and Holt strolled around the Rotunda and talked quietly about cases they had worked, men captured, and, in Siddler's case, promotions missed.

The walkie-talkie made routine reports: all quiet at the Paulson home; no activity at McVey Funeral Home. The night dragged on.

The visitors to the Rotunda had thinned to a trickle, then stopped. Some Capitol police stood here and there; the candles flickered; the honor guard of enlisted men kept their vigil under the watchful eye of a master sergeant.

At four-twenty, the radio beeped that McVey's had been asked to pick up a body in New York and that a hearse would be dispatched immediately.

"It's a phony," Holt said. "Follow it. Call Maryland State Police and ask them to assist. When they pick it up, drop off."

"Roger," said the dispatcher.

"Move Unit Two from the District line immediately to the funeral home," Holt said. "I want that gap closed—now."

"Roger."

"It'll take him about three minutes to get there," Holt said to Siddler. Then, into the radio, "Tell him to keep a sharp lookout for traffic coming the other way."

"Roger."

"Haverman still asleep?" Holt asked.

The dispatcher reported, "Roger on the colonel. No activity at the Paulson home."

Some minutes later, the dispatcher said, "Unit Two reports no traffic from the funeral home. He is on station. No activity from funeral home. State police report they can intercept New York-bound hearse vicinity of Bowie."

"Tell them to tag along for a while," Siddler said. "Let's retrieve Unit One, as soon as possible."

"Roger."

"Want to wake Gilmore and get him down here?" Siddler asked.

"No. Let him sleep. It doesn't mean much yet. Haverman hasn't moved. The hearse is moving away from here, not toward us. And there's no other activity. I might have been wrong."

A half-hour later, the dispatcher reported that state police had picked up the hearse and were trailing it north on Interstate 95 toward Delaware. Unit One was ordered back to the parlor.

"I'm going to take a break," Siddler said. It was five o'clock. The Rotunda was deserted.

He went quietly into the Statuary Hall for a smoke. He lit up and then wandered around, staring at the odd collection of statues of long-forgotten statesmen, otherwise obscure fellows in antique garb, who had served their country and their states with enough distinction to be memorialized in stone near the seat of power. Maybe Herb Paulson will make it here someday, Siddler thought. Someday far in the future. He found he liked it in the Statuary Hall, standing among incongruous stone men who kept their stern, earnest expressions despite the lapse of time and the people's memory. He thought, in a few hours the charade of this funeral will be over and I will fly back to Asia and resume my inconsequential duties.

A tall black soldier in full dress ceremonial uniform approached him. "You can't smoke in here, sir," the soldier said.

"Well, I am, cap'n," Siddler said. He turned to the man, who was smiling. Siddler felt a movement behind him and reached instinctively toward his .45. The soldier's hand closed over his wrist.

Siddler struggled. The soldier gathered him in with strong arms. Siddler turned in time to see Haverman putting a long-barreled pistol against his head.

"Good-bye, partner," Haverman said pleasantly.

"You—!" Siddler twisted frantically.

"Phfutt!" said Haverman's gun and Siddler fell down like a rubber man.

They were working on his face the way they had worked on Luckett's face. They had a crowbar and were hitting him with it, the pain gouging down through his skull, pulsating in two brilliant particles of light that made him want to vomit.

Something wheezed. Something heavy pressed on his chest. Sometime later, he pulled himself out of a pit of darkness. It was gray above. He pulled some more, floating up, giddy, nauseated.

Something stared at him. He squinted against the pain. He was looking at himself in a small mirror. The light was bad. But he could see himself. Then he sighed. He understood.

He twisted away, remembering the muzzle of the gun against his head, feeling the cold round O of the barrel. The bullet had gone into his neck somehow. He didn't know where else it might have gone. He hurt mightily, but he was alive. He wondered what the hell had happened to Holt.

Siddler stared at his watch. It was five-thirty. A lifetime in thirty minutes. He eased himself against the door that stood there. It swung open. Gray light came in. He had been lying in a washroom off Statuary Hall. The marble figure of some forgotten statesman peered at him sternly, the face he had seen through the door.

Heavy marble columns surrounded him. He guided himself among them. Pale dawn lit the Rotunda, a soft wash of gray hues touching the long murals around the walls. He walked hesitantly out through the colonnade and out under the huge, circular vault of the dome.

The master sergeant was quietly directing his men in removing the candelabra and other ceremonial articles from around the bier. The casket of Richard Paulson was gone.

18

Croak, said Siddler. Croak, croak.

The sergeant turned, his face opening in shock.

Siddler fumbled out his identification and moved forward. He tried to say words. The sergeant listened carefully and then said:

"Colonel Haverman, the senior honor escort, ordered the casket taken to Arlington Cemetery Amphitheatre about twenty minutes ago, sir. He had a hearse and three men from the Fort Myer honor guard detail with him . . . Yes sir, there was another man, a civilian, kind of heavyset, going right along with them. Wore a blue suit . . ."

The Ford custom rocketed along Independence Avenue, tires slapping heavily on the ancient trolley tracks. Gilmore swore a string of oaths. Siddler huddled in the back seat, dazedly taking in the scenery, his mind groping for sweet reason and finding nothing but dull ache everywhere.

The car bulled around the Tidal Basin and into the looping roads that skirt West Potomac Park, then up around the Lincoln Memorial, which glowed in light browns and pinks in the morning light. The car went up on Memorial Bridge, the wide, ceremonial arch spanning the Potomac, the honor route for the nation's warriors on their last march. They sped for the burial ground.

The cemetery mounted before them—a vast dark expanse of hillsides shrouded with trees, marked by rows and rows of white gravestones and larger tombs, pale ghosts in the dawn. The hills rose one behind the other. Near the top a small flame flickered and danced—the Kennedy flame. At the crest of the highest hill, the floodlit, marble-columned facade of Arlington House, once owned by Robert E. Lee, floated like a beacon. Gilmore spoke.

"No shooting. All we want is to get them and let the funeral go on as if nothing's happened. We're headed for the Tomb of the Unknowns, up near the Lee mansion. Behind the tomb is the amphitheatre. The national ceremony is to be there. Right under the amphitheatre is a guardhouse for the soldiers who march at the Tomb. The stage is almost directly above the guardhouse, hidden from view from either the guardhouse or the Tomb."

The sedan rolled around the wide circle on the Virginia side of the bridge and through the unguarded main entrance. A sign instructed: "All Visitors Turn Left—Hours 8 A.M.—8 P.M."

Gilmore parked the car on a grass fringe near a high hedge and they got out. A heavy, humid silence enveloped them, broken by the chirps of a single bird. Gilmore started up the hill and was soon well into the trees.

Siddler straggled after, his head throbbing, his brain a mosaic of light and shadow. He called upon the years of rich food, the years of booze, all the good times he had ever had, and prodded himself along.

The .45 hung like a rock in his right hand. He passed under a large, redstone arch that said:

Rest on, embalmed and sainted dead, dear as the blood ye gave.
No impious footsteps here shall tread the heritage of your grave.

The air was suddenly filled with the pealing of chimes from the gray hillside above. They were playing "Fairest Lord Jesus." The two men continued straight up the hillside, cutting across the paved pathways that meandered around the hill. They passed a small gold and blue sign:

(1) PLEASE ACT WITH PROPRIETY HERE AS ELSEWHERE IN THIS CEMETERY IN ORDER THAT OUR DEAD BE PROPERLY HONORED. (2) WHEN FUNERALS ARE IN PROCESS PLEASE DO NOT INVADE THE PRIVACY OF THE CEREMONY. (3) PLEASE WALK ON DESIGNATED WALKWAYS ONLY.

They went on a little farther and came to a parking lot. Across it, steps led up a small hillside to the Tomb of the Unknowns itself, a colonnaded marble plaza. They skirted the parking lot, keeping to the trees and foliage that surrounded it, then came to the bottom of the steps leading up to the Tomb. There was a steady cadence of sounds as the guard at the top walked his ceremonial patrol. Tap tap tap tap tap tap tap tap stop. Quarter-turn. Click of heels. Silence. Slap of gloved hands on rifle. Clashing of metal parts and leather as the weapon shifted shoulders. Tap tap tap tap tap tap tap tap stop. Quarter-turn. Click of heels. Silence. With clockwork precision, the honor guard performed his duties in the deserted cemetery.

"They do that for a half-hour at a time," Gilmore whispered. "Then they get relieved by a buddy from the guardhouse under the amphitheatre. We're going to let that fellow go to the other end of his patrol, then we're going up and over, past the Tomb, and to the amphitheatre. It's just beyond."

When the guard reached the other end of his march, Gilmore and Siddler mounted the steps and disappeared over the top. They crossed a small lawn between the Tomb and the amphitheatre, went up three short steps of the amphitheatre, and peered in.

The structure was of white marble—a perfect replica of a Greek amphitheatre. A colonnaded walkway circled the area and the amphitheatre itself fell away in a series of circular marble benches that descended toward a small stage.

Engraved in marble high above the stage were the words: "When we assumed the soldier we did not lay aside the citizen."

A heavy wooden coffin, trimmed in brass, rested on the stage. It was open. Haverman and Big Nick stood at the coffin, bent over it working furiously, pulling out plastic packets and dropping them into two canvas tote bags. Two black men in full dress Army uniforms stood on either side of the stage, cradling pistols. Robert Holt, his hands tied behind him, stood near Haverman.

Gilmore and Siddler retreated cautiously. "We have to get help," Gilmore pleaded, his face twisted with rage and torment.

Siddler shook his head slowly. The neck wound flared. "No time," he whispered.

"What'll we do?"

"Ambush 'em."

"They'll kill Holt."

"You can't stop it at this range."

"Jesus!"

"Shut up!" Siddler hissed. He looked around. A hearse stood fifty yards away in a grove of trees. Its rear door gaped open. Siddler jerked his head toward it . . . "Their wheels."

He started crawling away from the amphitheatre. Gilmore came along reluctantly. When they were far enough away, they got to their feet and went quickly toward the hearse.

Gilmore lay on the ground twenty feet from the hearse. He was concealed by a screen of azalea and rhododendron, about fifteen feet from the path Haverman and Nick would take to the hearse. The ground was cool, smelling sharply of tanbark mulch and freshly cut grass.

Siddler was hidden in the front seat of the hearse. They controlled the getaway. Surprise was with them. Siddler would fire from the hearse. Gilmore would fire from behind. They should be able to pull it off. Gilmore's heart pounded in his head. His targets were Haverman and one of the two soldiers.

After a time, he saw the five men coming toward him. The two fake soldiers came first, cautiously peering here and there like birddogs. Holt walked between Nick Westley and Rupert Haverman. Gilmore felt his guts tightening. The .38 fitted perfectly into his hand. Its hammer was back on full cock. All he had to do was brush the trigger and the pistol would fire, sending lead into . . . which target? He waited for a clue as to who should get it first.

They came along more slowly than he had thought they would. He waited, sensing tension among them. Both Haverman and Nick had their pistols out. Shoot one and the other shoots Holt. Shoot the soldiers, and Haverman and Nick shoot Holt. Holt was a sitting duck. He walked as if he knew it, disjointed, distracted, his face pale.

They were arguing, Gilmore suddenly realized. "I want it and I want it now!" a furious voice suddenly burst out. Nick was scowling in rage. Haverman carried both tote bags of heroin. How had that happened? It didn't matter. He possessed them. His pistol waved lightly in the air.

"When we're safe," he said. "Really safe. Then you get yours."

"I want it now!" Nick snarled. His gun barrel steadied on Haverman.

"Hold it," Haverman said. He backed away, shielding himself with Holt. His long-barreled pistol aimed at Nick's midriff and then waved the two black soldiers toward him. "Get right over there," he said.

Gilmore watched in agony. Siddler was too far away to get a decent shot. He couldn't risk it all himself. It would be the end of Bob Holt.

"We need each other," Haverman said coolly. "I've come too far and faced too much to botch this deal now. Nick, I've got the scag. I want to split it with you. You understand that?"

Westley nodded slowly, his pistol starting to move in a circle.

"The agent shields me, Nick. You can't get me. I'll blow you away. I'm an expert marksman. I never miss! I can't miss at this range!"

Nick hesitated. Haverman seemed to sense his indecision. "I'm going to walk toward the hearse with this narc between us," Haverman said. "I'm telling you ahead of time so you'll know exactly what I'm doing. I'm going to throw the bags in the back and close the door. I'll sit on the passenger side. Then your two friends can throw away their weapons." As he spoke, Haverman moved toward the hearse, keeping Holt between him and the black men. Gilmore could not shoot; Holt was in the way.

"When they've tossed their weapons, they can come aboard. You can bring your gun. I have no quarrel with you, Nick. I just want my fair share. Do you believe me?"

No answer.

"Do you?"

"Okay," Westley muttered. He lowered his pistol. Haverman backed up to the hearse.

Siddler saw the group moving toward him. They had taken a line slightly to the left of where he thought they would come. The hearse was parked so that he could not get a clear shot at them from the driver's side. He slumped down and peered through the rear window, out through the carrying space with its rich lining of velvet and satin and its frosted side windows.

Considering his options, he could see that Haverman had backed away from the others, using Holt as a shield. The colonel was fifteen feet away, one hand pulling Holt by the neck, the other on the long-barreled pistol. Gilmore had no clear shot.

Haverman was dressed in his full ceremonial rig, flashing dark blue tunic with brass buttons, lighter blue trousers with a wide stripe down them. His colonel's eagles glinted richly at his shoulders.

Siddler pulled his head down. His heart pounded. Haverman was too far away for a good clean shot. His shaking hands would never permit accuracy. He must wait. The moments dragged on. Then he felt a movement in the hearse.

Haverman was standing right at the hearse, half-turned toward the compartment, holding Holt in front with his pistol hand. He was swinging a bag into the carrying space. Siddler aimed. As he fired Haverman saw him, his face a mixture of surprise and horror. The shot boomed in the driver's compartment, shattering the rear window.

Haverman stumbled back, firing into the hearse. Two bullets plowed into the seat. There was another report as Nick Westley shot Robert Holt. Holt crumpled, a sudden deadweight. Haverman released him and dashed for the bushes. Gilmore suddenly emerged, .38 blasting, the bullets leaping into Big Nick. Nick flipped over with a scream. The two soldiers flung themselves to the ground as Gilmore ran to Holt.

"Bob! Bob!" He threw himself on Holt, rolled him over. "He's still breathing!" He shouted to the soldiers, "Get up! I need help!"

In the distance, other soldiers were running toward the hearse. Siddler crawled out and went around to the rear door. Holt was ashen-faced. He had taken a bullet high up in the chest. Siddler gauged the distance and got down on his knees. After a time, he found what he was looking for. Several large drops of blood lay on the grass where he had last seen Haverman. A gut shot. Slow and painful. Poor bastard. The blood led away in a faint trail. He could hear sirens wailing in the quiet morning. Siddler got to his feet and started after Rupert Kaiser Haverman.

The trail went past the Tomb of the Unknowns at a good distance, then down a small valley filled with serene white gravestones. Siddler crested the hill. There was a small sound nearby, like a cough. Siddler crouched as turf jumped a foot away. "Haverman!" Siddler whispered. "Give up!" There was no answer.

Siddler picked up the trail again. He found Haverman's uniform tunic. The dark blue jacket had a hole slightly to the right of the fifth brass button from the top. It was heavily soaked with blood.

Siddler moved up a hill, his strength waning. This was not the way it should be, he thought. There should be more. He was far from the hearse now, far from the sound of sirens. He and Haverman were alone in the cemetery.

He suddenly realized the trail had skirted well to the east and then come back to the west and he was now looking down at the flat, circular shapes of gray marble of the Kennedy gravesite, set into a hill below the Lee mansion. Traffic moved slowly across the Memorial Bridge toward the Lincoln Memorial in the distance; the day's first airplane was wheeling low over the Potomac on the way to National Airport downriver; a thin film of yellowish-brown haze was settling over the city, the day's dose of smog. But here it was peaceful and quiet.

Siddler followed the trail toward the Kennedy gravesite. He came up on the low, circular rampart along the heavily traveled path to the eternal flame, and peered across. Rupert Haverman was sitting with his back to the parapet, his feet out in front of him. He was clutching his middle and wincing. The pistol with the silencer lay on the path a few feet from him.

As Siddler advanced, Haverman seemed not to notice. He was breathing heavily. Siddler reached him and stood looking down, his .45 aimed at Haverman's face. Haverman was propped against a slab that said:

> *With a good conscience our only sure reward,*
> *With History the final judge of our deeds,*
> *Let us go forth to lead the land we love,*
> *Asking His blessing and His help,*
> *But knowing that here on earth*
> *God's work must truly be our own.*

Haverman slowly looked up at Siddler. "I thought I killed you."

"So did I," Siddler nodded.

"Help me."

Siddler shook his head. "I got some questions first."

19

Possible permanent damage, the doctor had said. So they put Siddler on an operating table and swabbed out the wound and took a look. Miracle, the doctor had said, the bullet went right through the neck muscles, much pain, but no serious damage. Missed the spinal cord by a half-inch. Day of observation required, the doctor had said. At which point Siddler struggled to his feet, whispering curses. So there was a compromise. They would let him go if he would agree to some precautionary treatment. He had agreed. Dumb move.

Blood poisoning, the doctor had said. So they gave Siddler a tetanus shot. Infection, the doctor had said. So they gave him a penicillin shot. So now, Siddler reflected unhappily, I am sore in the butt and the arm as well as the neck. Not only can I not speak, I also cannot write and I cannot sit. I lounge. Altogether, it is not a bad fate. People have wished me dead today and I am still here. They are dead. He swilled the bloody mary and watched Anita cruise around the apartment in her bare feet. She was nervous, waiting for Gilmore, he could see that. She moved like a cat. He found it disheartening. He made her nervous. He drank some more vodka and juice.

The telephone rang. Anita ran to it as though he would race her for it, and clung to the receiver. She hung up with a radiant smile. "Bob's going to be all right . . . he's going to live."

Siddler smiled and nodded. He jiggled his glass and raised his eyebrows, tipping his head at the empty tumbler. She jumped up and went for a refill. She was well trained. That made him think of Gilmore. And Holt. They had a nice thing between them. Older brother, younger brother. Uncle, nephew. Not quite father, son. He settled on uncle, nephew. Not much maybe, in the face of the bureaucratic crap they had to put up with. But more than he had. More than that.

After a while, Gilmore came in. He looked better. He had calmed down and seemed in control. Siddler stuck his hand out. They shook. Siddler smiled. Gilmore had made the right moves. It had turned out all right.

Gilmore mixed himself a rum and tonic. The girl had a Dubonnet on the rocks, with a twist of lemon. Siddler had another bloody mary. It was an elixir. Soon, he knew, he would be able to talk. Gilmore took out a tape recorder and set it up and began talking. "Today is June 22," he said and gave the year and identified himself and Siddler. Then he said, "It is now five o'clock in the afternoon and the time has come for a summation of the events of recent weeks, which culminated today in the national ceremony for Richard Herbert Paulson."

He described picking up Luckett's trail, moved to General Tran, banker Pei and to Haverman, the penetration of the ring that smuggled Typhoon heroin into Washington, beginning with the first shipments and carrying through to Siddler's arrival in the U.S. with Haverman. He brought it forward to the shootout.

"There is other information which should be told at this point," he said. He turned the recorder off. "Tell what you

know that you didn't hear from me so far." Siddler nodded
and held the mike near his lips.

"All this from interview with Rupert Kaiser Haverman:

"Haverman met Nick in Saigon in April. Nick had come
there to set up a connection. Haverman embittered, wanted
money, action. They worked it out: used Manes in Tansonnhut. Shipped heroin-laden bodies. Passed info via Cronder, who was forced to cooperate by Nick, to McVey Funeral Home, Nick's uncle. Nick had man at Dover Air Base
morgue for unpacking, done chiefly after hours. Used mostly
shot-up bodies—those not suitable for viewing. Whistler
was courier supplied by Haverman to insure safe passage
for corpses. He made seven trips. Haverman thinks Whistler engineered intercept of Samuels body last week at
Hickam or Travis, kept heroin for self. He is not sure.

"Haverman believed self endangered when Luckett body
mixed up. Told Nick, arranged large final transfer, speeded
up when spotted me. Thought self safe when Watson chopper burned. But returned to Saigon, couldn't find Manes,
alarmed at that. Saw Paulson angle and realized its good
cover. Designated self personal escort, diverted transfer
case to Shed Sixteen area where loaded heroin himself,
caught next CONUS freedom bird. When I caught up in
Hono, was spotted by Haverman and took only safe step,
joined him in media spotlight. Haverman and me washed
in media, press coverage, both safe as result, neither
dare risk move against other. Publicity shielded Haverman,
made impossible unmasking for fear of damage to country,
powerful senator, so forth. We both grasped that. It tied us
to each other."

Siddler paused and gargled a mary. It soothed his throat.
Then he said: "Haverman disbelieved my arrest by Gilmore
and Holt at Andrews. Very worried. Decided to wait until

positive no tail, no tap, no bug. Took chance on Paulson house. Ordered flowers sent to McVey home, sent message therein. At nightfall, Nick over fence, makes call, arranges for early Paulson body snatch. Haverman out Paulson window, elude known tail. McVey decoys narcs, Nick borrows friendly hearse. Creep to Capitol. Nick as soldier. Get me. Go to Rotunda, Haverman walks up to Holt big as day, engages conversation, points gun, has hostage, orders body moved . . ."

His voice had grown ragged. He turned away from the mike. "Goddam lie detector," he whispered.

Gilmore took over and quickly described the chase and gun battle near the Memorial Amphitheatre. Then he said, "What did you, Mr. Siddler, do, after you found Colonel Haverman?"

"Questioned him."

"And then what?"

"Killed the sombitch."

Gilmore said: "Soldiers and police and rescue personnel arriving at the scene of the shooting had no knowledge that either Mr. Siddler or Colonel Haverman had taken part in it; they did not know of the interrogation of Haverman by Siddler; they did not know of the shooting of Haverman by Siddler. At the amphitheatre, we found that the conspirators had removed all trace that the casket was raided."

"I used his own gun," Siddler whispered. "Silencer."

Gilmore continued:

"I was interrogated by Arlington County Police Inspector George K. Nedwin at the scene. I did not know the whereabouts of Haverman and Siddler at that point and simply excluded them from my account. Nedwin believed very little of what I told him and insisted on repeating numerous questions. Finally I broke off discussions with

Nedwin and telephoned Noel Walker to inform him briefly of the shootings and ask for instructions. Walker assured me he would plug any holes regarding Rupert Haverman until we discovered what the colonel's fate was.

"I then informed Nedwin that he would have to speak with my superior, Walker, for further information. I used with him the cautionary warning we have used with other agencies in pursuing this affair—I told him that the incident was a matter of national security. I could not respond to his questions, particularly those relating to whether the hearse was the same one that had transported the lieutenant from the Rotunda to the amphitheatre area.

"Nedwin was not satisfied with this, but there was little he could do. The cemetery is a federal reservation within Arlington County and primary police responsibility rests with the federal government, which has jurisdiction over crimes committed on its lands. Nedwin, although obviously eager to get into what clearly was a sensational case, was on shaky ground. He finally released me. He later conferred with Noel Walker at Walker's request.

"Siddler returned to the amphitheatre sometime later and was immediately hospitalized. He refused any but minimal treatment. Haverman's body was found at eight-oh-two A.M. by a tourist from White River Junction, Vermont, who was seeking a better camera angle for a photograph.

"Mr. Siddler and I subsequently were debriefed by Mr. Walker in the presence of a legal stenographer, a member of the White House counsel's staff, and the principal assistant U.S. prosecutor for the federal District Court of Washington. I understand that the original of the stenographic record of that session is in the personal safe of Mr. Walker and that copies have been supplied to the Defense Department and the FBI.

"Questions arose during the session about the pertinence of making public the entire information regarding the Typhoon shipments. It is now time for the "Six O'Clock Evening News," a nationwide television network news show. The tape record of this disclosure will contain an audio soundtrack of the show, which reports on daily events of interest in the country, the capital, and the world."

Gilmore turned the television on and set the tape mike up in front.

"Very cute," muttered Siddler. "What're you going to do with the tape?"

"I think I'll put it in the family archives."

An even-featured newsman appeared. "Good evening. Two stories occupied the hearts and minds of the nation's capital today. One was an event of solemn dignity as a young war hero was buried with full military honors at Arlington Cemetery after a national ceremony in which many of the capital's leaders sought to evoke a new spirit of cooperation and dedication in the land.

"The other was a mysterious tragedy, according to police and Administration sources, as the honorary escort for the young war hero apparently inadvertently wandered into a vicious gun battle between federal narcotics agents of Task Force Washington and a group of black drug dealers who were attempting to make off with a record one hundred ten pounds of high-grade heroin. The escort, Colonel Rupert K. Haverman, apparently was gunned down by an escaping dope dealer as he was seeking to aid the federal authorities. More on these stories in a moment. But first . . ."

Siddler had another drink during the commercial. He was beginning to feel especially good as the vodka whirred away inside him.

He watched the TV with interest as a series of scenes

from Arlington Cemetery appeared—a ceremony at the amphitheatre, Senator Paulson addressing the crowd, then a formal burial service with enlisted honor guard, the folding of the flag, the gun salute, and "Taps." The last scene was of the grieving parents. The image faded, and was replaced by a large still photo of Haverman—his military file photo —and a smaller, blurred photo of Nick Westley.

"Combat veteran, escort, new friend of Senator Paulson," said the newsman, identifying Haverman. "Drug dealer, criminal, heroin smuggler," the man continued, identifying Westley.

"Somehow, the war hero and the drug dealer crossed paths early today in Arlington Cemetery as Westley was completing a heroin transfer that law enforcement sources say may be worth between twenty and forty million dollars. When the encounter was over, the colonel had been slain. Federal agents killed Westley and took into custody two henchmen and the heroin. One federal agent was seriously injured. Federal authorities, obviously embarrassed that this shooting occurred so near the place where Richard Paulson was to be interred, clamped a tight lid of security on the entire affair, answering few questions and leaving much unsaid.

"However, this much could be learned from a high Administration source. Police theorize that Colonel Haverman, not wishing to disturb the Paulson family, slipped from the senator's house in the early morning hours of today and went for a visit to the Memorial Amphitheatre and Arlington Cemetery where his young charge soon was to be laid to rest. While there, Colonel Haverman apparently blundered into a bizarre shootout between federal agents and a gang of dope peddlers active in the Washington area.

"Police believe that Haverman attempted to aid the federal agents, but was spotted by one of the suspects and shot. The colonel, who was armed with his personal sidearm, a .38 caliber automatic pistol, sought to escape. He had been shot in the abdomen by a suspect armed with a .45, police say. The colonel got as far as the Kennedy memorial area, where he was shot down in cold blood with his own weapon. A Pentagon spokesman has confirmed rumors that Colonel Haverman's pistol was equipped to handle a silencer. They say this is not unusual for some types of missions in Vietnam battle areas. The colonel's pistol did not have a silencer attached when it was found.

"One unexplained aspect was a report that the remains of Lt. Paulson were moved from the Capitol at least two hours earlier than scheduled by a military officer some sources said was actually Colonel Haverman himself. The sergeant of the honor guard at the Rotunda was not available for comment on this puzzling part of the case.

"Federal sources reported that the hearse found at the scene was there for another, unrelated funeral ceremony and that any damage to the vehicle would be repaid by the government.

"The federal account leaves much unanswered. Did the smugglers habitually use the national cemetery for their operations? What other suspects are there? Why were Nick Westley and two henchmen dressed in Army uniforms? Was there some plot on the part of the smugglers to discredit the ceremony, and if so, why won't the government talk? And finally, what of the movements of Colonel Haverman in his last hours? We intend to press these and other questions in coming nights on Network News.

"Meanwhile, the White House issued a statement deploring the incident, and calling Colonel Haverman 'a credit to his country.' The White House said plans are being made for stepped-up security at the cemetery . . ."

Gilmore flipped off the television and the recorder. Silence filled the room.

"Well, well," Siddler said at last, "I guess they found it . . . not suitable for viewing. How about another mary?"

Epilogue

In December of 1972 in an obscure magistrate's hearing room in federal court in Baltimore, a young assistant U.S. attorney recounted a fantastic tale. An international conspiracy of heroin smugglers, he said, had for eight years been bringing multi-kilogram quantities of high-grade heroin into the United States from Southeast Asia inside the dead bodies of American soldiers being shipped home for burial.

The prosecutor elaborated. The conspiracy was ruthless and sophisticated. Its members penetrated the complex military transport system by impersonating soldiers. The case before the magistrate involved a man caught wearing an Army sergeant's uniform, complete with battle ribbons and other official insignia. The man carried a green military identification card and mimeographed travel orders. He was taken off a Military Air Command flight from Bangkok, Thailand, after FBI agents diverted the flight to Andrews Air Force Base near Washington, D.C. The man had never served in the U.S. military in his life.

Aboard the flight with the fake sergeant were two coffins containing the bodies of dead American fighting men from Indochina, according to the federal attorney. The bodies had recently been cut open and stitched shut again. Acting on information from a tipster, agents opened the bodies, but they found no heroin. Later, they speculated that the heroin had been removed by members of the conspiracy during the flight's twenty-four-hour layover in Honolulu.

The fake sergeant was convicted of impersonation and sentenced to twenty years in prison. Despite its courtroom assertions and despite a police and legal battle against other suspects of the alleged conspiracy that continues today, the government never found heroin being shipped inside any soldier's body.

K. K. and P. McC.
Washington and Saigon
October, 1973